Globesity

Globesity
A Planet Out of Control?

FRANCIS DELPEUCH,
BERNARD MAIRE,
EMMANUEL MONNIER and
MICHELLE HOLDSWORTH

publishing for a sustainable future

London • Washington, DC

First published by Earthscan in the UK and USA in 2009
Reprinted 2010

This book is a revised and updated version of an authorized translation of the work *Tous Obeses?* by Francis Delpeuch, Bernard Maire and Emmanuel Monnier, published by Dunod (Paris, 2006). Translation by Lorna Scott Fox.

ISBN: 978-1-84407-667-3 paperback

Typeset by FiSH Books Ltd.
Cover design by Rob Watts

For a full list of publications please contact:

Earthscan
Dunstan House
14a St Cross St
London, EC1N 8XA, UK
Tel: +44 (0)20 7841 1930
Fax: +44 (0)20 7242 1474
Email: earthinfo@earthscan.co.uk
Web: **www.earthscan.co.uk**

Earthscan publishes in association with the International Institute for Environment and Development

A catalogue record for this book is available from the British Library

Library of Congress Cataloging-in-Publication Data
Tous obeses? English
   Globesity: a planet out of control? / Francis Delpeuch...[et al.].
      p. cm.
   Includes bibliographical references and index.
   ISBN 978-1-84407-666-6 (hardback) – ISBN 978-1-84407-667-3 (pbk.)
   1. Obesity–Popular works. I. Delpeuch, Francis. II. Title
   RC628.T6813 2009
   362.196'398–dc22

                                                                                2008048033

At Earthscan we strive to minimize our environmental impacts and carbon footprint through reducing waste, recycling and offsetting our CO_2 emissions, including those created through publication of this book. For more details of our environmental policy, see www.earthscan.co.uk.

This book was printed in the UK by MPG Books Ltd, an ISO 14000 accredited company. The paper used is FSC certified and the inks are vegetable based.

Mixed Sources
Product group from well-managed forests and other controlled sources
www.fsc.org Cert no. SA-COC-1565
© 1996 Forest Stewardship Council

Contents

List of Figures and Boxes

Figures

Boxes

Foreword
Human Health and Global Crisis

By the year 2000, there were more adults with obesity than adults who were underweight, taking the world as a whole. The rapid rise in the numbers of adults and – especially – children who are overweight and obese is unprecedented, and the potential costs in terms of ill-health, personal incapacity and lost economic productivity have been described as a 'public health crisis', as a 'time-bomb' and as a 'tsunami'.

The economic and agricultural development that has brought about this extraordinary change in human health prospects is remarkable for its success in terms of supplying the calories needed to feed so many people – but it is absolutely unsustainable. As presently designed, the global system for food production, processing and distribution fails to meet our needs. It fails in several ways: it fails to ensure good nutritious foods are well distributed, it fails to ensure that healthy foods are cheap – at least as cheap as unhealthy foods – and it fails to ensure that the production of food can continue into the future.

At present, food is very poorly distributed in many regions of the world, especially sub-Saharan Africa, but even in Europe some people enjoy access to shops overflowing with food from all over the globe while others must endure days when they eat only one meal or no meal at all. The cheapest food available in most developed economies is the least healthful: it consists of 'empty' calories (starches, sugars, fats and oils, in the form of soft drinks, snack foods, confectionery), persuasively designed, packaged and promoted. The production of this food depends on large amounts of non-renewable energy sources, and the pollution from this production is contributing significantly to global warming and the creation of arid deserts.

From basic principles we can state:

- Human populations depend on healthy ecosystems
- Where populations have lost their local ecological integrity – as in urban areas – they depend on healthy ecosystems elsewhere to support them

- The demands on the world's ecosphere already exceed its capacity and therefore cannot be sustained.

The present book addresses these issues and shows how closely interwoven are the forces that shape our dietary health and our environmental and economic security. It recognizes that world history is at a point of transition, and that there is hope that we can yet avoid future disaster. The current crisis brings together all of us concerned with protection of the environment, maintenance of biodiversity, promotion of public health and the struggle for social justice and human rights.

Tim Lobstein
Director of Projects and Policy
International Association for the Study of Obesity, UK

Preface
Tragic Blindness

Conquer hunger at all costs. That was the titanic challenge faced by the newly created international bodies, in particular the Food and Agricultural Organization of the United Nations (FAO), after the Second World War; and at that time the battle looked far from being won. Not only in the countries that would later be called 'Third World', but also in Europe as it struggled to get up from the rubble, the needs of a population that was growing by some tens of millions every year were huge. And the second half of the 20th century was to encounter more than its fair share of shortages and famines. From New York to Paris, from São Paulo to Kolkata, the goal was therefore clear: to produce as much food as possible, at the lowest possible cost. And to give everybody at long last the right to a full stomach; even more, for the inhabitants of developed countries, the guarantee of enjoying meat every day of the week.

This long-term campaign was conducted with undeniable efficiency. In Europe, eating turnips was soon just a bad memory, while mass production of sugar, oil and cereals silenced the Cassandras who once predicted widespread famine by the turn of the millennium. The spectre of global shortfalls that haunted our minds a few decades ago has faded away in the light of more optimistic predictions. In 2003, the FAO world report on the state of food and agriculture declared that 'After 50 years of modernization, world agriculture production today is more than sufficient to feed 6 billion human beings adequately.'

Nevertheless, the great paradox of this new abundance is the fact that 850 million people worldwide are still going hungry today, and suffering from diseases caused by the lack of essential nutrients. Recurrent food crises, such as that which struck Niger in 2005, continue to afflict entire populations. Young children in many countries are disproportionately exposed to malnutrition.

But if our first priority remains to feed the starving, it is also true that many international organizations have failed – or were unwilling – to wake up to another crisis, one which has lately leapt to prominence: the masses of this planet are not only suffering from hunger and want, but also from a calorie surplus. Shocking? Certainly. Unexpected, at any rate, by world decision makers. At the Millennium Summit, the most urgent goal agreed by UN member states was the eradication of extreme poverty and hunger around the world; not a line was written about chronic disease, the category to which obesity and its consequences belong.

In the same way, when Lula da Silva became president of Brazil in June 2003, his anti-poverty programme was encapsulated in the simple slogan of 'Zero Hunger', for like his electorate he took it for granted that the fight against poverty consisted above all in providing more calories. The tragic blindness of this assumption was shatteringly revealed, in late 2004, by a study of family nutrition from the Brazilian Institute of Geography and Statistics. Among its stark findings was that the prevalence of overweight in the adult population was now ten times greater than that of underweight.

Too much fixation upon the notorious excesses of the United States, that land of large people force-fed on ice-creams, obscured the fact that close to 40 million Brazilians – or 40 per cent of the population, including children and adolescents – were too large themselves.

Does this mean that standards of living are rising for Brazilians across the board? The truth is crueller, because it is a mistake to equate obesity with affluence. There's no doubt that the rich were the first to pile on the pounds. But they have since been overtaken by the poor, whose lack of education and resources is making them progressively heavier than any other group. The facts contradict all our preconceptions: there are already more obese individuals in developing or newly industrialized countries than there are in the industrialized world. Before long, China may well wrest from the US the unenviable title of the 'fattest' nation on Earth.

Another paradox is that to be plump is not the same as to be well fed. Although there are enough calories available, vitamins and minerals are often lacking. Thus the poorer districts in emerging countries have the sad privilege of seeing chubby teenagers and growth-retarded young children playing side by side in the street. Obesity does not respect borders, whether these are geographical, social or economic.

The French considered themselves to be shielded from the epidemic by solid culinary traditions. The age-old cuisine, coupled with the benefits of red wine – the famous 'French paradox' – would act as an impregnable Maginot Line. But this illusion has been shattered by a less glorious reality. Today we realize with horror that even French children, in the not too distant future, may well be able to hold their own in the weight stakes against any of their American peers. And we are beginning to realize that no society is immune from this plague, which is rapidly engulfing the whole of the globe.

In short, there is no time to be lost. For those extra pounds are not just a matter of aesthetics: fat kills! It kills 80,000 people a year in Brazil alone, where some cities have now banned the sale of confectionery and soft drinks in schools. France took similar steps in 2005. But of course nobody is so ingenuous as to believe, on either side of the Atlantic, that such measures will be enough to do the trick. Obesity differs from most other diseases in that it springs from more than one cause. It thrives on a multitude of factors, all of which compound together to create a globally 'obesogenic' environment.[1]

The impulse to identify specific culprits is a powerful one, nonetheless. Many have succumbed to the temptation, pointing the finger at fast-food culture and the creeping 'McDonaldization' of the world. However well founded this accusation might be, it should not be allowed to absolve the many other actors in the drama, from farmers to consumers, including the supermarkets, the media and even the town planners who design the way our cities work; all these have played their part. How far does the real responsibility of each one of them extend? Over and above the fist-waving, there are too few scientific certainties as yet to allow good and bad marks to be awarded, let alone to justify blaming any factor in particular. An even more difficult task will be for everyone to agree on a set of solutions acceptable to all. For obesity is not so much the illness of an individual, no matter how greedy that person may be; it is the illness of the world that is feeding its hunger. And we will never overcome it until we are prepared to rethink, in depth, the ways in which we produce, sell and consume our food. It would therefore be absurd and hypocritical to blame the overeaters themselves: they are merely the product, albeit perhaps too efficient and too well adapted, of our greater collective choices.

Note

1   Obesogenic environments are defined as 'the sum of influences that the surroundings, opportunities or conditions of life have on promoting obesity in individuals or populations'. Concept introduced by Boyd Swinburn of Deakin University, Australia.

List of Acronyms and Abbreviations

AFSSA	Agence Française de Sécurité Sanitaire des Aliments
ANIA	Association Nationale des Industries Alimentaires
BMI	body mass index
BMJ	British Medical Journal
BSE	bovine spongiform encephalopathy
CAP	Common Agricultural Policy
CARHCO	Central American Retail Holding Company
CERTU	Centre d'étude sur les réseaux, les transports et l'urbanisme
CO_2	carbon dioxide
CH_4	methane
CLA	conjugated linoleic acid
EAGGF	European Agricultural Guidance and Guarantee Fund
EASO	European Association for the Study of Obesity
EFSA	European Food Safety Authority
EPODE	Ensemble, Prévenons l'Obésité des Enfants
EU	European Union
FAO	Food and Agriculture Organization of the United Nations
FDA	Food and Drug Administration
FSA	Food Standards Authority
GMO	genetically modified organism
HWWI	Hamburg Institute of International Economics
IOTF	International Obesity Task Force
INPES	Institut National de Prévention et d'Éducation pour la Santé
INRA	Institut National de la Recherche Agronomique
IRD	Institut de Recherche pour le Développement
NCD	non-communicable disease
NICE	National Institute of Health and Clinical Excellence
N_2O	nitrous oxide

PAL   physical activity level
PNNS   Programme National Nutrition Santé
PorGrow   Policy Options for Responding to the Growing Challenge
             of Obesity
SCN   Standing Committee on Nutrition (UN)
SDC   Sustainable Development Commission
SOS   Swedish Obese Subjects
UN   United Nations
vCJD   variant Creutzfeldt-Jakob disease
WHO   World Health Organization

Chapter 1

Wave of Panic Across the Planet

In March 2004, and again in October 2005, the US House of Representatives approved a bill that may one day be seen as emblematic: The Personal Responsibility in Food Consumption Act. The purpose of this Act was to protect the food and agriculture industry from being sued by obese consumers. It sought to relieve producers, retailers and distributors from any responsibility at all for the expanded waistlines of their over-faithful customers, and more importantly, from liability for any damage wrought by overweight to those consumers' health. In future, no such 'frivolous' lawsuits (as they were officially dubbed) would be admissible.

Whether in irony or clairvoyance, the text rapidly became known as the 'Cheeseburger Bill', given how blatantly it looked out for the interests of the fast-food industry, whose importance to the US – not least in terms of jobs – is huge. According to the sponsors of this bill, citizens bingeing on burgers and fries have no one but themselves to blame for the consequences. Overeating is an individual choice that does not fall within the remit of the law; still less does it implicate the corporations that produce and market this deluge of food.

It's basically a matter of 'common sense and personal responsibility', proclaimed the original proponent of the bill, Republican congressman for Florida, Ric Keller, in 2004. His argument was immediately echoed by the spokespersons of McDonald's, which welcomed with manifest delight the removal of any justification for using the company as a scapegoat before the law.

Indeed, Ronald McDonald, the kiddies' friend, had been getting increasingly on the nerves of many older consumers. In 2002, two teenage girls from New York filed for damages against the celebrated 'restaurant' chain on grounds that it was liable for their obesity, their diabetes and their high blood pressure. Other consumers were quick to jump into the ring. More and more lawsuits were being brought against the home of the Golden Arches, as people demanded compensation for having wrecked their bodies with 'McJunk'.

At first, these cases were largely ridiculed by the general public. Look, the gluttons taking it out on the larder – what a topsy-turvy world! Mainstream moralism, always disapproving of self-indulgence, was shocked. Of course, the debate around obesity never fails, now as then, to wind up the advocates of virtue and discipline who can't stand people setting themselves up as victims, when plainly their only problem is lack of willpower. Why can't they just face up to their addictions!

Everyone has their own ideas and prejudices on this point. But the fast-food industry and other dispensers of junk food, bottom-of-the-range crisps, sweets and soft drinks, did not for a moment underestimate the danger represented by a cascade of individual and group lawsuits against them, as spectacular and costly as those that had recently engulfed the tobacco industry. For American justice, to give it its due, came down very heavily indeed upon the smoke merchants. In September 2004 they were sued for the astronomical sum of US$280 billion, the highest claim ever submitted to a civil court, on the grounds that the industry had knowingly deceived the public about the inherent risks of smoking. In the event, the judges were not convinced by the federal government's case, and the suit was dismissed in February 2005. The defendants could breathe again, but it had been a close shave. Especially when we remember that back in 1998, the same industry had already come to an arrangement with 46 states of the Union whereby the states agreed to halt proceedings in return for the tidy sum of $206 billion, to be paid over 25 years.

So might there be an analogy between tobacco and fast food? Ominously for the calorie peddlers, people who suffer from obesity and diabetes are beginning to see one. The federal government itself has been known to make the comparison between them, as when the Health and Human Services Department Secretary, Tommy Thompson, declared to a press conference in March 2004 that 'We need to tackle America's weight issues as aggressively as we are addressing smoking and tobacco.' The tone was ringing and virile, quite at odds with the same government's past efforts behind the scenes to discredit the recommendations of a string of experts from the World Health Organization (WHO). Their advice clearly spelled out the direct correlation between bad diets – those rich in calories, fats, sugar and salt, but poor in vitamins and nutrients – and the rise of obesity, with all its attendant health risks. Such conclusions had always been fiercely rejected by the Bush administration, and so perhaps Thompson's speech signalled that the government was beginning to recognize the scale and seriousness of the problem. But it's a long way from talk to action, naturally. Meanwhile the 'Cheeseburger Bill' lives on, to remind any doubters out there that the very official 'crusade' to fight the flab will pull any punches that might hurt the US economy. And yet... A succession of books and reports all sharply critical of the food and agriculture business, including a hard-hitting book by the American nutritionist Marion Nestle (2007), are making their mark regardless. In the current climate, producers and vendors of substandard food are aware of the tide turning, a sea-change that could conceivably wash away the tasty profits they have made from our cravings.

Surfing this new wave the US film director Morgan Spurlock, used to making programmes designed to shock, produced an explosive documentary about the nation's diet, 2004's surprise hit, *Super Size Me*. The idea behind the film was to chart, in TV-reality style, the physical deterioration of a healthy person who is seduced by a restaurant chain's standard trick of offering much larger portions for a fractionally higher price. After all, when the difference is no more than a few cents, who can resist the lure of a double ration of fries or a mega-container of popcorn, who can say no to 2 fizzing litres of a certain brownish soda, up from a stingy 1.5? The marketing boffins know perfectly well that in the absence of genuine hunger, the vision of what seems like a 'bargain' can always be relied upon to get the saliva flowing.

So for a whole month, the initially lanky Morgan Spurlock, a New York yuppie of 33, ate all his meals at McDonald's, cheerfully munching through every dish on the menu. Whenever there was a Super Size option to be had, he had it. He underwent a complete medical examination before embarking on this uninhibited diet, and was monitored throughout the experiment, as a precaution, by a team of specialists including a GP, a gastroenterologist, a cardiologist and a dietitian. The results were even more alarming than one might have expected. Spurlock filmed himself growing flabbier and more pot-bellied by the day, while his complexion went to pot. By the end of the month the kamikaze filmmaker had put on 11 kilos (nearly 2 stones), that is, an average of 1 kilo every three days. And these effects were a trifle compared to the unseen ravages within. His cholesterol was stratospheric, his liver was on the blink and his partner was growing pretty tired of his non-performance in bed.

The demonstration was a crude one, no doubt, but it hit the target with a bulls-eye. The hamburger giant's PR teams deplored the suicidal irresponsibility of Spurlock's behaviour for all they were worth, but the damage was done: the film had planted an indelible message in the public mind, which now associated going to McDonald's with getting fat and possibly endangering one's health.

It was surely a coincidence when the fast-food chain withdrew all its Super Size offers just days after the documentary was premiered, at the Sundance festival of independent cinema in January 2004. Salads and fresh fruit are now prominently displayed on the menu. This is a positive development, even if the majority of customers are still going in for burgers and fries, which remain the company's most lucrative products.

In France, anti-'McDo' demonstrations have long been standard fare. Today, however, voices are being raised with unprecedented vigour, calling for a move beyond symbolic speeches and actions onto the terrain of tough measures to safeguard sound principles of food and nutrition. It was in this spirit that a Public Health Act was passed, in September 2005, to ban the presence of snack and fizzy drink vending machines on school premises. And it was more than a token gesture, because it highlighted the perennial problem of the responsibility for the spread of obesity of the food conglomerates by the kinds of foods they promote. And the heightened tone of the arguments over even this limited measure says much about the stakes that are being played for here, at the crossroads of health and economics.

Meanwhile, in March 2005, the French Member of Parliament Jean-Marie Le Guen presented a bill of his own aimed at tackling 'the obesity epidemic'. Among the measures it envisaged was 30 minutes of compulsory exercise, every day, for every child in school; installation of free drinking-water fountains in schools, annual weight checks for children and the creation of a top-level commission dedicated to the fight against obesity, charged with making sure that convenience or processed foods stick to the rules governing maximum percentages of sugars, fats and salt. Brands that overstepped the limits would be fined.

Shocking figures

Why the fuss, and why now? It's hardly a scoop to reveal that many Americans eat too much unhealthy food and are grossly overweight. News and magazine articles, TV reports and box-office movie hits such as Shallow Hal, with a fat-suited Gwyneth Paltrow, have long been obsessed with what is also the first thing to strike any regular visitor to the US: every ten years or so, the country's average measurements goes up a size. This has been the case ever since the early 1980s, when the phenomenon suddenly rocketed (weight gain among Americans had remained moderate from 1960 to 1980). So marked was the change, according to dismaying data published by the International Obesity Task Force (IOTF), that the number of obese people more than doubled in 25 years. The condition is currently reported to affect almost one-third of adults and half of all African-American women. Close on two-thirds of all Americans are already overweight, meaning that the day may not be far off when the population of certain states will be entirely made up of 'well-upholstered' citizens.[1]

Box 1.1 Where obesity begins

How is obesity defined? Historically, the term was coined by the medical profession to denote cases in which the excess of body fat is such that it interferes with an individual's physical and mental health. But what exactly were the thresholds? How was this fatty tissue to be evaluated, what formula would be simple and at the same time reliable enough to serve the purposes of vast statistical research projects? Epidemiologists eventually adopted the body mass index, or BMI, defined as the individual's body weight (in kg) divided by the square of their height (in metres).

$$BMI = \frac{weight\ (kg)}{height^2\ (m^2)}$$

Since 1997, the WHO has defined adult overweight as a BMI of 25kg/m^2 or over, this number being the cut-off point above which mortality begins to increase, whatever the morphology of the individual. Obesity proper is considered to set in with a BMI of 30, and becomes 'morbid' or 'massive' –

implying serious health hazards – above 40. It is an irony of history that the index was originally proposed, during the 1980s, as an index for identifying critically underweight people in undernourished populations.

It would be a mistake, however, to regard obesity as an American curse. The French, for example, have a long tradition of condescension toward what they see as a nation of bloated Big Mac addicts, and their mockery of Americans as hopelessly fat – a stereotype that still persists – conceals the subtext that they've got what they deserve, leaving the rest of the world feeling unconcerned by the antics of a society that has made a trademark of excess. We have all heard the stories, and they are true, of American travel companies such as Southwest Airlines deciding to charge obese passengers the price of two seats, or coffin builders[2] having to adjust their standard measurements to the dimensions of this new market.

By complacently deploring the bad habits of Uncle Sam, we have allowed ourselves to overlook a far more disturbing state of affairs, whose reality has finally come home to us with a force that is all the more shocking: the obesity epidemic has spread to the whole planet, and nothing seems about to stem its relentless progress. In 2000, the WHO already estimated the numbers of the obese at some 300 million, while over a billion were overweight. Indeed a recent review by Kelly and colleagues (2008) agreed, with even more pessimistic results for the spread of obesity, as they found that 396 million adults were obese and 937 million overweight in 2005.

A sizeable proportion of these people do indeed live in the US. But many can also be found in Canada, a country that is less frequently cited in this context but where the figures do not lag far behind; many others live in Eastern Europe, or in the Near East, where the rise of obesity that sometimes rivals the US has long gone unnoticed, due to the lack of available studies. Kelly and colleagues go on to predict the even worse news that by 2030, more than half of the world's population will be overweight or obese.

Europe in turn would be wrong to think itself immune, for the evidence suggests that it is already helplessly following the US down the same slippery slope. 'The prevalence of obesity has increased by about 10–40%

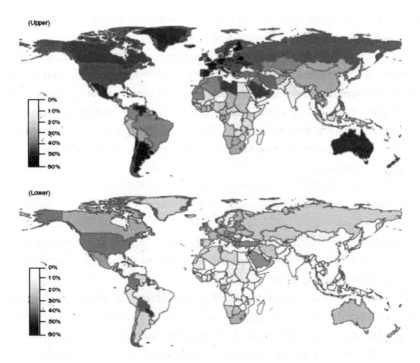

Figure 1.1 Worldwide prevalence of obesity (lower)
and overweight (upper) in 2005

Source: Kelly et al (2008)

in the majority of European countries in the past 10 years' warned the
WHO in 2000. In other words, several countries within the EU present
rates of obesity that vie with those found across the Atlantic. Like it or not,
we have no choice but to face up to the numbers: current data reveal that
in Cyprus, the Czech Republic, Finland, Germany, Greece, Malta and
Slovakia, the proportion of overweight adults is actually higher than in the
US – where it is already worryingly high, as we have seen. Three-quarters
of German men over the age of 25 are overweight. The most alarming
increase has occurred in Britain, where obesity prevalence literally doubled
between 1980 and 1995. Nine other European countries present an adult
obesity of over 20 per cent – which is to say, more than one in five adults![3]
Other developed countries around the world are hardly doing much

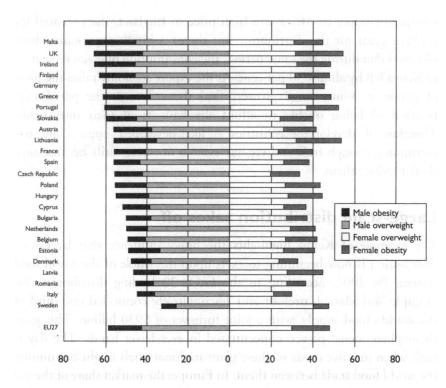

Figure 1.2 Adult obesity and overweight in the European Union

Source: IOTF data from www.iotf.org/database/index.asp accessed September 2008

better. In Australia and New Zealand, 10 to 15 per cent of the population is obese, according to WHO estimates, and in the United Arab Emirates, as many as one in three married women fall into this category. Japan, traditionally the epitome of frugality, has held out for longer than most – but it seems that here, too, they're catching up with the trend. The national census compiled in 2000 revealed that 27 per cent of men over the age of 20, and 21 per cent of women, were overweight, forcing the National Institute for Health and Nutrition to admit that obesity had become a public health issue for Japan as a whole.

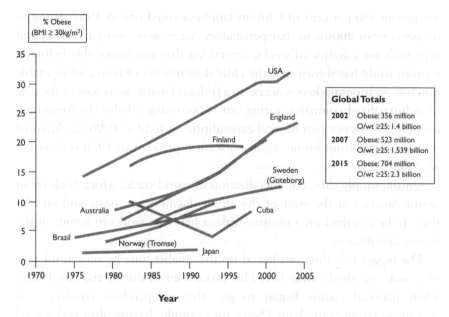

Figure 1.3 Global trends in obesity prevalence

Source: James (2008)

Developing countries are not exempt

Is obesity the inevitable scourge of spoiled nations? That cliché is well out of date. There are already more obese people in developing and newly industrialized countries than there are in the industrialized world. Here is one of the most startling paradoxes of a disease that for too long has been regarded as merely the downside of abundance: it is spreading like wildfire through poor countries as well. In September 2005, the WHO found that over three-quarters of all women were overweight in some 20 countries. Predictably, the US was on the list – but so were South Africa, Jamaica, Jordan and Nicaragua. In Mauritius, the proportion of obese men between the ages of 25 and 74 rose from 3.4 per cent in 1987 to 5.3 per cent in 1992, but the proportion of obese women soared during the same period, from 10.4 to 15.2 per cent. In other words, female obesity in that age bracket escalated by 50 per cent in just five years! At that rate, the number of obese Mauritians is set to double every ten years or so. Again according to the WHO, one Indian woman in five is currently overweight. In Delhi, obesity affects only 1 per cent of men and 5 per cent of women in the poorest neighbourhoods, but in middle-class districts, it has reached a staggering 32 per cent and 52 per cent, respectively. China is also beginning to make its weight felt; in this as in other domains… But the world record belongs to a handful of Pacific islands. One of them is Samoa, where epidemiologists have found, to their alarm, that over three-quarters of urban-dwelling women are obese; the men score almost as badly at around 60 per cent. Sounding a note of caution, the WHO reminds us here that for an equal body size, Pacific Islanders have a lower percentage of body fat than subjects of Caucasian origin. Their real obesity threshold may thus be higher than the standard BMI would have it, because the cut-offs between BMI ranges have been defined on the basis of criteria extrapolated from white populations.

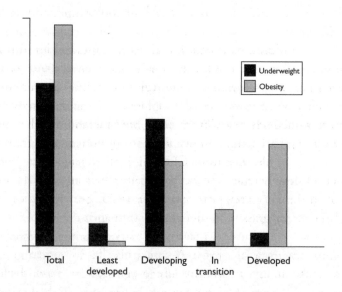

Figure 1.4 Adult population affected by underweight and obesity according to level of development

Source: WHO (2000)

Box 1.2 The limits of numbers

BMI, calculated as a height to weight ratio, makes no distinction between lean body mass (made up of bones, muscles, etc.) and fat body mass. Depending on body size, the same BMI may thus correspond to different proportions of fat. There is clearly no comparison between a well-developed, muscle-bound athlete and a paunchy couch potato, and yet both may register the same BMI. Another factor is that women tend to have a higher proportion of body fat than men with the same BMI. A similar difference may be found across ethnic groups. Polynesians, for example, typically display a lower degree of fatness than white Australians with an equivalent BMI. Conversely, some Asian groups (especially in India) have a higher proportion of body fat for the same BMI. For these reasons researchers are cautious about comparing different populations. It is generally more useful to study the overall fluctuations of BMI within a single population over a period of time.

The end of the 'French Exception'

So what is the outlook in France? *French Women Don't Get Fat*, boasted the title of the 2004 book by Mireille Guiliano that became an instant best-seller in the US. And it's true that France is still doing pretty well in this respect. French rates of obesity and overweight are among the lowest in Europe – especially among women, of whom less than 30 per cent experience weight problems (half of the percentage of the US!). An Obepi survey in 2006 (Charles et al, 2008) reckoned that 11.9 per cent of people aged 15 and above were obese, while almost a third were overweight. This means that over 40 per cent of French adults have a weight problem, either mild or severe. The figure is not negligible of course, but it easily undercuts the percentage of overweight in certain other countries.

Does this entitle the French to boast that the war has been won, singing the praises of French culinary genius? Unfortunately not. The much-touted *'exception française'* (what others might call French chauvinism) fooled many for a long time into thinking they could escape the epidemic. If the Germans were getting fat, that was only to be expected. If the Spaniards and the Greeks were following suit, well, too bad for them. But it could never happen to the French, because France – like a certain Gaulish village defended by Asterix – would resist the occupation to the bitter end; because good food is never fattening. The truth of the matter is crueller. While it remains the case that French adults are slimmer – relatively speaking – than their roly-poly neighbours, French children are certainly making up for it, and at terrific speed, for they have already closed the 'gap' and today they weigh in on a par with the average child in other European countries.

Children in the front line

It is the obesity among children and adolescents everywhere that has got so many epidemiologists worried. The figures are ominous, to say the least. The first wake-up call, as so often, came from the US, where the proportion of overweight children doubled between 1975 and 1995, from 15 to 30 per cent. Then, almost a decade later in May 2004, Philip James – president of the IOTF – announced that the obesity epidemic among European children had now spun 'out of control'! The data bear him out. In England, the rise in numbers of overweight children from 15 to 30 per cent occurred between 1995 and 2005, that is, it happened twice as fast as in the US. Faced with this ticking bomb, in June 2004 the Parliamentary Health Commission energetically sounded the alarm: 'It is staggering to realise that on present trends half of all children in England in 2020 could be obese', thundered David Hinchliffe, Head of the Commission. Only a few years behind, Poland seems to be following the same accelerated pattern. And the rest of Europe has nothing to be proud of, either. Its children are putting on the pounds at a rate that can increasingly compete with the standard set by the US. Mediterranean countries are among the worse hit, so that in Spain, Italy, Albania or Greece, we find the numbers of overweight children already climbing to between 30 and 40 per cent. In 1986, 23 per cent of Spanish six-year-olds needed to shed some puppy fat; ten years later, 35 per cent of them did, or more than one in three.

This is an area in which France has no more lessons to teach anyone. With almost 20 per cent of children aged 7 to 11 classified as 'too fat', France is placed right in the middle of the European pack. So why has the '*exception française*' lost its protective powers over its children? It is difficult to say. But there is no denying that the way people live has changed profoundly, in France as elsewhere, in the space of one generation. Any remaining differences between the lifestyles of French and American youngsters seem to be vanishing fast, in a one-way direction. However, this may be changing, as recent data from France suggest that childhood obesity may be falling slightly, although this is restricted to higher socioeconomic groups. Other countries are also finding that childhood obesity rates are abating (Sweden and Switzerland) or reaching a plateau – as is the case in the US, where there was no increase between 2003 and 2006. Writing in the *British Medical Journal* (2008), Tim Lobstein, Director of the Child Obesity programme at the IOTF calls for cautious

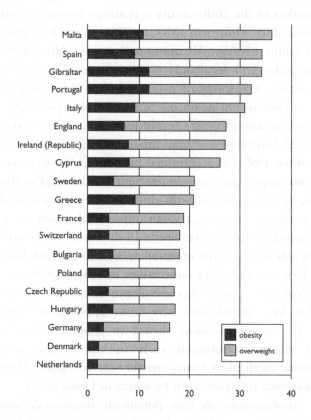

Figure 1.5 Estimated percentages of children aged 7–11 obese or overweight for selected European countries.

Source: IOTF (2005), www.iotf.org/database/index.asp

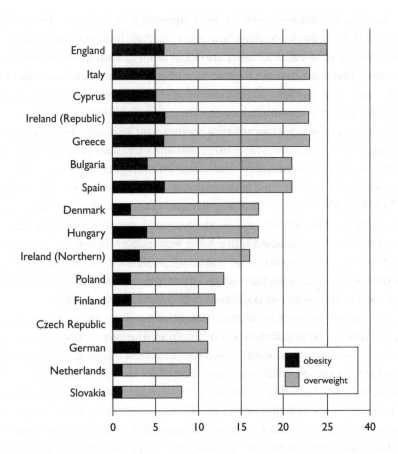

Figure 1.6 Estimated percentages of children aged 13–17 obese or overweight for selected European countries

Source: IOTF (2005), www.iotf.org/database/index.asp

optimism about whether this really represents a sea-change: 'Whether the tide is really starting to turn is hard to judge.'

Whatever the case may be, child obesity is steadily establishing itself across the globe. In Japan, the number of obese 6 to 14 year olds is thought to have doubled between 1974 and 1994, when it reached 10 per cent. In Thailand, 12.2 per cent of children aged 6 to 12 living in Bangkok were obese in 1991, rising to 15.6 per cent in 1993, which represents a 3 per

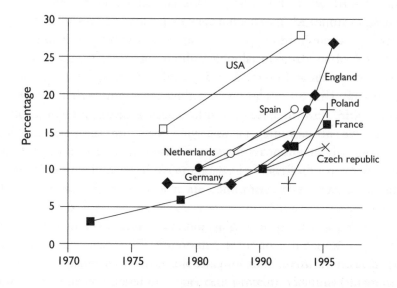

Figure 1.7 Rising overweight in children

Source: IOTF (2004), www.iotf.org/database/index.asp

cent hike in just three years. The same trend has been observed in Saudi Arabia, where a study found an obesity rate of 15.8 per cent among boys aged between 6 and 18 in 1996.

Too poor to be thin

This tide of obesity among children and adolescents is rising especially fast in low-income households. There is a bitter paradox here, for the rule seems to be that the heaviest people are the ones with the lightest wallets. This law now holds true almost all over the world. For it turns out that once past a certain level of economic development, it is the poorest and least educated who tend to become obese the fastest. Only the very poorest inhabitants of the poorest countries offer an exception to the fatal formula, and that is because they lack the resources to provide for even basic needs, and are severely undernourished in most cases.

So is obesity, like malnutrition, an illness of deprivation? The suggestion may seem incongruous at first. And yet we cannot fail to acknowledge that calories are no longer beyond the reach of even modest budgets; what's expensive these days is a varied diet. Without enough money to buy fruit and vegetables, struggling households stock up on sugar, starch, oil and other processed foods – high energy and low cost. Greasy meals can make one feel full for next to nothing. And who has the heart to deny the kids a few cheap sweets, when they are already deprived, for want of cash, of the toys and activities enjoyed by their schoolmates or those fun-loving teens of their favourite TV series. This is a common phenomenon in the US, where the pockets of greatest obesity are the same as those of poverty. It does not apply so much in France, for example, where the inverse correlation of weight to income is far less marked than in most other countries (educational levels appear to exert more influence than income by itself). But the correlation remains significant nonetheless,[4] especially among youngsters. Slimness is becoming an unattainable luxury for the poorer families in our midst.

Notes

1   Some studies indicate, however, that the obesity rate is levelling out in the US.
2   At Goliath Casket, speciality manufacturers of oversize coffins, triple-wide models are in increasing demand.
3   European Association for the Study of Obesity (EASO).
4   4.6 per cent of men and 3 per cent of women under the age of 75 earning more than €1500 per month qualify as obese, compared to 11.4 per cent and 9.3 per cent, respectively, who earn less than €900, according to the French survey *Baromètre nutrition-santé* 2002.

Chapter 2

'Badnosh' and Other Paradoxes of the Abundant Society

The relentless advance of obesity sends out an unmistakable message: most of us are putting away far more food than we really need. But is there any reason, if you'll forgive the pun, to 'make a meal' of this fact? There certainly is, because the problem goes much deeper. Nutritionists have been telling us for years that we eat too much, and, more importantly, that we eat all the wrong things. Food-lovers have registered their own complaints, helping to popularize a telling concept that is all the rage in France: the notion of '*malbouffe*', loosely translatable into English as 'badnosh'. The aptness of this term propelled food journalist Jean-Pierre Coffe to media stardom, as with stinging Rabelaisian verve he proceeded to make mincemeat of such monstrosities as limp lettuce and tasteless tomatoes. But he also convened panels of experts who, albeit in more conciliatory language, came to much the same conclusions: there is definitely something amiss in our kitchens. Consumers themselves, meanwhile, have become thoroughly confused about the whole issue. They fall eagerly upon anything that labels itself 'traditional' or a return to 'old-fashioned home cooking', and yet they are not, deep down, convinced by the slogans. They are uncomfortably aware that they don't know what they're eating any more, and this anxiety came violently to a head during the early 1990s, with the outbreak of mad cow disease.

Cows go mad

It all started rather quietly. In April 1985, a veterinary officer came to examine a cow on a farm in Kent in England. The animal was behaving

Box 2.1 What is *malbouffe*?

The term '*malbouffe*' has a popular ring to it, and yet it was first employed in 1979 by a scientist, Joël de Rosnay, then director of the Cité des Sciences in Paris. Its use is now so commonplace in France that it appears in the most formal dictionaries. In 2001, the Petit Larousse defined it as 'poor diet having a negative impact upon health'. The Hachette dictionary nails it more narrowly as 'food that is unsatisfactory on both the nutritional and the gastronomic planes', a definition that might open the door to some lively discussion, since nothing is more personal than tastes about taste.

oddly: it was hyperactive, aggressive toward the farmer, and seemingly unable to control its legs. One year later, bovine spongiform encephalopathy, or BSE, was detected in nine herds in England. This was still only a sickness to which cows, in rare cases, were prone; it was a headache for a handful of farmers and vets. A mere five years later, the entire British beef industry had collapsed, leaving the British people – soon followed by other Europeans – nursing a deep scepticism towards any public statements affirming the safety of the food they ate.

For the disease spread at an amazing speed. By 1993, more than 37,000 cows had gone incurably 'mad' in their turn and had to be culled, to the despair of the traumatized farmers (some of whom took their own lives). But the nightmare did not end there. In 1995, a new form of neurodegenerative disease, similar to BSE, was identified – but in a human organism this time. This variant of Creutzfeldt-Jakob disease, with the acronym vCJD, soon became a household word. For it is a particularly terrible affliction: it literally eats away at the sufferer's brain, riddling it into a sponge-like structure. The unfortunate victims were thought to have contracted the disease by eating infected beef. There was panic up and down the land. The media raised the spectre of thousands, possibly even hundreds of thousands of deaths to come.

Horrified consumers discovered that their Sunday roast had been nourished on a feed manufactured from the offal and bones of its fellow cattle, as a means of disposing of these. Cows forced into cannibalism! The image of peaceful herds grazing in verdant meadows took quite a knock. Welcome to the world of intensive livestock farming, with its assembly-line productivity, requiring it to minimize costs while delivering maximum quantities of meat. Who cared about a little disrespect to Nature? Hitherto, consumers had been far too thrilled by the cheapness of fresh meat to worry about the gloomy, concrete-floored hangars where chickens lived jammed into battery cages, or where sows had their young in breeding stalls too narrow for them to turn around in. But now, perhaps, the public was finding out about the sordid downside of cheap food.

Today it is thought that the epidemic was caused by a peculiar type of protein, know as prions. These are mutant proteins, equipped with the formidable ability to make 'normal' proteins adopt the same pathological form, and thus, in a sense, to multiply themselves. These prions, originally appearing in a cow, are assumed to have been transmitted from one bovine to the next through the feed that incorporated meal from infected carcasses.

This kind of feed had formerly undergone heating at such high temperatures than any prions were destroyed. In the name of reducing costs, however, it was decided during the 1980s to reduce the temperatures. This turned out to be a serious miscalculation, a thrifty attempt to cut corners that brought an entire industry to the brink of ruin.

The crisis was so brutal that the whole system crashed, practically overnight. Consumers, producers, distributors – nobody trusted anybody else. The consumer accused the butcher, the butcher blamed the cattle farmer, and the farmer raged against the Ministry of Agriculture that had done nothing to protect him.

In 1997, Britain's Labour government tried to heal the trauma by creating the Food Standards Agency, which began operations in 2000, charged with making sure that food was safe to eat. However, any genuine overhaul of agricultural policies, that might be adequate to addressing the systemic causes of the crisis, proved more difficult to set in motion.

In France, the equivalent Agence Française de Sécurité Sanitaire des Aliments (AFSSA) was created in 1999. Other watchdogs of this kind were founded all over Europe. The year 2002 saw the birth of the European Food Safety Authority (EFSA) to underwrite the health status of food products at continent-wide level.

The consequences of the mad cow crisis were profound, and its repercussions are still with us today. For the first time, a major revolt had challenged the basis of 40 years of productivist agriculture, with values such as food quality and consumer health. Even if sales of beef cuts have climbed back to pre-BSE levels (beef is currently the most popular meat in France), consumers have become irrevocably aware of the importance of what they are eating – and also of their total helplessness at the hands of the food production system. They have seen through the rustic version of where our food originates, in busy farmyards where proud cockerels greet the rising sun among kind, aproned countryfolk. They now know that agriculture had moved on without their knowledge, to leave them at a much greater distance from the product than they had once thought.

Other parallel crises have erupted, frequently enough to keep the psychosis alive. The egg industry imploded when in 1988 the British Health Secretary admitted that salmonella was rife among poultry. Only a year later, the spotlight moved across the Channel as traces of a carcinogenic substance, dioxin, was found in Belgian flocks. The same disastrous script seemed to play out in every case: a significant risk to public health, the failure of experts to assess it with any degree of precision, the revelation of horribly unsavoury farming practices, and the reminder that it is impossible for us to know exactly where our food comes from. Time after time, consumers feel menaced by a looming crisis that they can do absolutely nothing about. And they wonder: what's left that's fit to eat?

GMOs: A focal point for new fears

Against this background, the issue of genetically modified organisms (GMOs) was very much a last straw, at least within Europe. Breaking with centuries of 'classical' selection techniques, which were based on repeated cross-breeding or hybridization, a new hi-tech school of agriculture went on to introduce, directly into an organism's genome, genes extracted from other living things – whether bacteria, viruses, plants or animals – in order to create certain desirable properties. Gigantic salmon, for example, or maize that is resistant to certain parasites. Genes may also be incorporated to act upon a natural gene by deactivating it, a procedure that can, for example, delay the ripening capacity of fruit.

The first genetically engineered crop, the slow-maturing McGregor Flavr-Savr® tomato, was created in the US in 1994. It was withdrawn from the market three years later. Since then, however, other transgenic crops have thrived in the US: herbicide-resistant soya, maize, cotton and

Box 2.2 An Indian trail leading to mad cow disease?

It is now a matter of consensus that mad cow disease, or BSE, was transmitted between animals through the insufficiently heated animal-based flours used in their feed. But how did the original cows get it? By a genetic mutation, or perhaps the leap into bovids of scrapie, a related disease that has long been known to affect sheep? In the 3 September 2005 issue of the British medical magazine *The Lancet*, Professor Alan Colchester put forward a more controversial theory. In his view, the Kentish cows may have gone 'mad' after eating animal-based meal whose ingredients originated in the Indian subcontinent. These flours had been manufactured from the remains of cattle, mixed, according to the author, with human remains infected with another strain of transmissible spongiform encephalopathy, close to Creutzfeldt-Jakob disease. This may not be as implausible as it sounds. During the 1960s and 1970s, Britain did indeed import hundreds of thousands of tonnes of bone and offal from India, Pakistan and Bangladesh, for use in making fertilizers and animal feed. In the countries concerned, peasants were paid to collect bones for the trade. It would have been quite possible to pick up human bones by mistake, for there are plenty of them about. Tradition dictates that a dead body must be incinerated and the ashes thrown into a river, but due to lack of firewood, many corpses are not properly burnt. Alan Colchester pointed out that the occurrence of human remains among animal bones exported from the subcontinent has several times been noted. The vector of a human encephalopathy present in some of these tissues could then have found its way into the bonemeal used to feed English calves, and this is how the first herds might have been infected. This theory has yet to be confirmed.

rapeseed, pest-resistant maize and cotton, and many more. The welcome in Europe, by contrast, has been icy to say the least. Up came the defences in Gaul, Germania, Iberia and elsewhere. Why the distrust? Because over and above the uncertainty about the environmental impact of GM foods, European consumers, thoroughly put off by earlier food scares, are flatly opposed to any biotech manipulations whose long-term consequences for human health remain largely unknown. The Old Continent put its faith in an old adage: when in doubt, don't.

Whether for good or bad, GM foods have been the hook on which to hang a range of contemporary terrors. These include: fear of multinational companies taking out a patent on life, rejection of ever more intensive farming practices, anxiety for the environment in terms of a possible cross-contamination between related species, fear of the appearance of new allergies and more. The French rural activist José Bové, with his warrior moustache and vivid turn of phrase, has used these fears as the motor for high-profile campaigns in favour of a model of 'peasant agriculture' that, superficially at least, chimes in with the more comforting image of husbandry that alienated consumers yearn to recover. Faced with the cruelties of industrial livestock production, tasteless tomatoes or strawberries and the distastefulness of hormone-pumped beef and dioxin-enhanced chicken, the calls for a shift towards healthier practices – for the sake of environmental as well as human health – are growing louder.

Too much fat and sugar, not enough taste

José Bové and other champions of the movement for a simpler, less processed diet have an irritating way, at times, of overstating their case. But each of them in their own fashion has drawn attention to an inescapable truth: the food we eat has undergone a wholesale transformation in just a few decades, without us really noticing. And the changes are not all for the better.

Let us quickly pass over the tasteless fruit and vegetables to be found in every supermarket. They are what you get from the peculiar logic of selecting and developing new varieties for their luscious appearance or capacity to survive packing and transport, at the expense of what is surely an essential quality: their taste. Oddly enough, the blandness of much of this produce is the result of a deliberate choice on the part of the growers, who wish to offer innocuous experiences that will appeal to as many people as possible. Out, then, with radishes that 'burn your mouth' or the slightly bitter kind of cauliflower – they have too much character, they're a minority taste, and in all fairness consumers can hardly blame the farmer for catering to their own preferences. But the food processing industry has become such an expert at dosing synthetic flavourings and colourings, that one can enjoy the illusion of a fruit yogurt when not a particle of fruit has been near it; yogurt mixed with real fruit is positively a let-down after that.

od culture when the taste of a product bears
to its components!

ith the so-called delicatessen counters, with
their displ ......s and reformed ham and other pork-like confections
of an unnatural pink, the reconstituted eggs shaped into long bars, making
it easier to slice them into rounds, the chemically crab-flavoured sticks, the
cocoa-free powders infused with heady 'chocolate flavour', the aseptic
pasteurized cheeses whose garish wrappers are hopeless proxies for the
absent savours within. Much has already been written about the loss of
culinary values and the gradual demise of a whole gastronomic culture.
But, just like the spectacular food scares that made such an impression
upon the public, this visible decline is only the tip of the iceberg.
Nutritionists are concerned about a less obvious development that has far
more serious implications for our health. Too much simple sugar and too
much fat, not enough fruit, fibre or vegetables: our diet is getting more
unsuitable year on year.

Home cooking

Over 70 per cent of EU citizens consider themselves to already eat a
healthy diet, according to a pan-EU survey conducted in the late 1990s.
The reality is less gratifying. Today's nutrition experts unanimously
describe as a balanced diet one in which at least 50 per cent of calories are
ingested in the form of carbohydrates (sugars, in all their guises), around
15 per cent in the form of proteins, and no more than 35 per cent in the
form of lipids, or fats. The Western diet as a whole – takes only 45 per cent
of its energy from carbohydrates, and around 40 per cent from lipids,
sometimes more. That is to say, there is an obvious surplus of fats, and a
shortage of carbohydrates. In addition, the carbohydrates we tend to
consume are increasingly composed, not of starch and fibres, but of
sucrose, fructose and glucose: monosaccharides, or simple sugars, which
are easily digested – and assimilated by the body.

It's the spoonful of table sugar stirred into one's morning coffee, it's the
fructose lurking in a fizzy drink (a litre of which contains on average the
equivalent of 17 sugar cubes!), as well as in sweets and pastries. Altogether,
these simple sugars provide 10 to 20 per cent of our total energy intake.
But they provide it in the form of what are called 'empty' calories – pure

Box 2.3 Fat and sugar: An ancient passion

Our love of sugar is ancient and powerful: if you wet a baby's lips with sweetened water, it starts smiling at once, which proves that the attraction to sugar is innate. Some evolutionary anthropologists even hold our immoderate craving for sweet things to be a factor that might have favoured the runaway development of the human brain (for the brain can only function by burning glucose, the simplest sugar there is). But what might have been a beneficial weakness way back at the dawn of the species, when sugar was extremely scarce, becomes a handicap in periods of abundance. The same goes for fats, to which we, like other animals, are irresistibly drawn. Their presence in food makes everything taste a good deal better. Besides, fat contains far more energy than an equivalent mass of other types of nourishment (9kcal per gram, as opposed to 4kcal per gram in proteins and carbohydrates). Thousands of years of shortages and even famines must no doubt have conditioned us to love energy-rich foods, and given us an appetite for luxury for which our bodies are now paying the inflated price.

energy, empty of the minerals, vitamins and other micronutrients needed by the body and plentifully supplied by less processed foods. To make matters worse, other carbohydrates such as cereals are commonly stripped down to their core elements, in an effort to remove the 'impurities' whose crucial input for our bodies is beginning to be grasped more and more clearly.

The great example of such wrongheadedness is the sad fate of our bread, which until very recently was expected to emulate a sheet just out of the washing machine and look 'whiter than white'. Most people are unaware that refined flours have been stripped of the magnesium enclosed in the original grain, so that recommended daily intakes of magnesium are not always being met. And it is probable that this kind of bread, which tastes of nothing and keeps so badly, has contributed to the fall from grace of a staple food that had always been uniquely symbolic of the diet, even in France. In any case, the upshot is that the French get through five times less bread today than they did at the start of the 20th century (165g per day compared to 900g, according to the French National Institute for Statistics and

Economic Studies). Perhaps it's also because of the persistent myth that bread is fattening. On the contrary, for the last 15 years nutritionists have been doing their utmost to rehabilitate a low-fat food that is full of nourishment, so long as the flour is not over-refined. The complex sugars it contains are assimilated slowly by the body and their ability to satiate tends to make us feel fuller sooner. And when bread is baked from wholemeal flour or flour enriched with bran and wheat-germ, it comes packed with roughage, that benefits the intestinal tract, as well as with vitamins and minerals, which are concentrated in the germ and the husk of the grain. In short, good bread is good for everyone, weight-watchers included.

Box 2.4 Joules and calories, two units of energy

The nemesis of slimmers, calories are units of energy measurement. Spelled with a small 'c', one calorie is the amount of energy necessary to heat a gram of water from 14.5°C to 15.5°C, under normal conditions of atmospheric pressure (101,323 pascal). However, dietitians often speak of Calories with a capital 'C', or kilocalories (kcal), a unit of one thousand calories – in other words, the amount of energy required to heat 1000 grams of water – or a litre – from 14.5°C to 15.5°C. A joule is the unit of energy employed by physicists. 1 calorie = 4.18 joules; so 1 kcal = 4180 joules (or 4.18 kilojoules).

Nutritionists are also working to promote other sources of complex carbohydrates, such as pulses. Split peas, beans, lentils and chickpeas have fallen out of fashion. And yet, unlike the sugary snacks and processed foods that have replaced them, these starches contain a diverse range of essential micronutrients. Any sugars they carry are fibre-coated, making them harder to digest, hindering or preventing their absorption into the body. Epidemiological surveys have also confirmed the benefits of eating large quantities of fruit and vegetables, in order to be protected from certain cancers, among other advantages. Such surveys have likewise highlighted the dangers of consuming excessive amounts of salt, for sodium raises the blood pressure and can lead over time to cardiovascular problems. This matters, because we are taking in far too much of it: about 10 to 15 per cent more, on average, than during the 1990s, probably because we have become that much fonder of ready meals, whose salt content is particularly high (80 per cent of salt intake derives from this type of food). In response, the French food safety agency (AFSSA) sounded the alarm in 2002, setting the target of a 20 per cent reduction of median salt intake within five years. Similar goals have been set throughout Europe, and the UK's Food Standards Agency recommends a 30 per cent reduction in intake, advising adults to take no more than 6g of salt a day, roughly the same as one teaspoon. Cutting down on salt is all the more necessary since potassium, naturally present in fruit and vegetables and an effective counterweight to excess sodium, remains deficient across the board.

Our health in jeopardy

Unsurprisingly, this ever more unbalanced diet is no good for our health. The cholesterol builds up, increasing the risk of cardiovascular problems. Dental caries are multiplying and some cancers have become more common, clearly as a reflection of certain diets. The WHO, which has monitored the steady progress of such pathologies all over the globe, is deeply preoccupied by the trend: 'Obesity can be seen as the first wave of a defined cluster of NCDs [non-communicable diseases] now observed in both developed and developing countries', notes a WHO report (2000). Viewed in this way, obesity is like the tree that prevents us from seeing the wood, the forest of diseases as varied as they are crippling, all of them connected more or less directly to what and how we eat.

The most spectacular of these diseases is without doubt type 2 diabetes, which has jumped to epidemic proportions in almost every part of the world. 150 million people were already estimated to be suffering from it in 2003 (WHO/FAO, 2003). This figure, which falls almost certainly on the conservative side, is set to double by 2025. Type 2 diabetes is directly responsible for 27,000 deaths per year in a country the size of the UK, tantamount to the population of a small town. Another worrying development is that while type 2 diabetes used to be called 'maturity onset', because it was an older person's disease, it now occurs earlier and earlier – we see it in teenagers, and even in children. This state of affairs would have been inconceivable 20 years ago.

Until recently, children and adolescents were principally at risk from type 1 diabetes. A rarer form, accounting for between 10 and 15 per cent of cases, type 1 is associated with genetic predisposition and is contracted when the autoimmune system mistakenly attacks the beta cells within the pancreas, in the region known as the islets of Langerhans. Beta cells are needed to produce the essential hormone, insulin, which enables our body to convert blood glucose into energy and to store the surplus. In the absence of insulin, the body becomes unable to process blood glucose, causing it to build up dangerously, and draws instead on its reserves of fat. The patient begins to lose weight – hence the sometime designation of this form as 'thin' diabetes – and feels constantly hungry and thirsty, with a frequent need to urinate, the urine containing high levels of sugar.

Type 2 diabetes is based on a completely different mechanism. It evolves very slowly, and presents no symptoms during the early stages. This means that many diabetics, exactly how many we do not know, are unaware of their condition. Type 2 was formerly known as 'non-insulin dependent' or 'fat' diabetes. It is most often triggered by an unhealthy high-calorie diet combined with low physical activity. In the first stage, insulin becomes unable to do its job; it continues to be secreted, yet fails to perform its regulatory action. Or in other words, its target, such as the muscles, ceases to respond to it. Such 'insulin resistance', which is unfortunately common among obese subjects, is due to the way muscles that are encased in fat gradually tend to draw on this plentiful reserve of energy, rather than using glucose. The liver, meanwhile, carries on producing glucose and releasing it into the bloodstream. And so the pancreas is stimulated into producing more, and increasing ineffectual, supplies of insulin.

Some researchers wonder whether insulin resistance may not be an adaptive mechanism deployed by the body in order to get rid of surplus fat, by burning it off inside the muscles. The procedure is initially successful. But where there is too much accumulated fat, the body rapidly finds itself overwhelmed.

The effects upon our health are often devastating. Once insulin has lost its efficacy, glucose can be neither stored nor consumed; it simply builds up in the bloodstream. In compensation, the body steps up the production of insulin so as to restore the balance. Ten or 20 years later, the pancreas that makes the hormone becomes worn out and throws up its hands in defeat. Insulin plummets. There is no longer anything to control the accumulation of blood sugar, and the excess glucose begins gradually to attack the sufferer's blood vessels, nerve tissue, kidneys and retina (diabetes is the chief cause of blindness in the US). It may also inhibit the healing of cuts and sores, provoking gangrene that can sometimes only be dealt with by amputation of a limb.

Type 2 diabetes is not always associated with obesity, and not all obese people become diabetic. But studies show that the obese are ten times more at risk from diabetes than individuals who are not. Worse still, obese women aged 30 to 55 are 40 times more likely to develop this type of the disease than their skinnier sisters! (WHO/FAO, 2003). And the greater the weight, the greater the risk.

Obesity brings other ailments in its wake. Vascular problems in particular, leading to heart attacks and strokes, especially when fat accumulates within the abdomen. In the US, the Second National Health and Nutrition Examination Survey has shown that overweight adults with hypertension were three times more numerous than normal weight adults with hypertension. The longer a person has been obese, the higher their risk of hypertension, especially among women. Other bad news about the consequences of obesity is for pregnant women – summarized by Hitchen (2007) in the *British Medical Journal*. It cites a UK report that found that obesity is now the key factor in why mothers die in pregnancy and childbirth – taking over from suicide in previous years. Deaths are usually indirectly related to cardiac causes due to the extra weight carried.

Obese people also have greater problems with breathing. This is because the fat clustered in and around the abdomen, ribs and diaphragm stiffens the ribcage, while fat lining the perimeter of the neck compresses the respiratory tract. These interferences, aside from causing stouter people to snore in an antisocial fashion, may trigger the 'obstructive sleep apnoeas' that plague over 10 per cent of the obese, both male and female: the sleeper's airflow is repeatedly blocked, as many as several hundred times a night in the most extreme cases. The sufferer stops breathing for 30 or 40 seconds, suffocating, unable to suck air into their lungs until the brain, alerted to the emergency, steps in to wake them up. This condition is obviously a disaster for the victim, who begins the day exhausted and spends the rest of it fighting off drowsiness. Over a quarter of people defined as massively obese (BMI>40) are prone to this form of sleep apnoea.

According to several studies, overweight is also a factor in certain types of cancer (including colorectal and gall bladder cancers, and, among post-menopausal women, endometric, ovarian, cervical and breast cancer;

among men, prostate cancer). However, it has not y~~et been mentioned~~ whether the risk factor for such cancers is the state ~~of obesity itself, or~~ rather the intake of certain foods in larger quantities t~~han others~~.

Obesity gives rise to a wide range of hormonal prob~~lems~~. We now know that the fat storage cells (called adipocytes) additionall~~y produce a number~~ of hormones, or very similar secretions, whose bu~~ild-up affects the~~ equilibrium of our body processes. An excess of fat m~~ay trigger the early~~ onset of puberty in a girl, and deregulate the ~~ovaries cycle in older~~ women, possibly leading to fertility problems.

Other unpleasant consequences of this con~~dition include~~ ~~gall~~stones, which appear three to four times more frequentl~~y in obese~~ people than in those of average weight (paradoxically, they are also very common among people who are losing weight). Finally, there is a heightened risk of osteoarthritis, since corpulence places great strain on the joints, giving a hard time especially to the knees.

To sum up: the more you are overweight, the sooner you die. The optimum size, for anyone who aspires to live to be a hundred, can be expressed as a BMI score of between 18 and 25. It seems probable that life expectancy, which has steadily risen in the developed world, may soon become shorter as a result of obesity. Unless we can overcome this epidemic, today's young people may well have unhealthier and shorter lives than their parents enjoyed before them.

Unhealthy bodies, unhappy minds

Physically unfit as they are, most 'fatties' are also damaged in terms of self-esteem. Let's face it: adiposity gets short shrift in our societies. This reflex may be acceptable or not, but the truth is that obesity is most often interpreted as the sign of a feeble character, and the obese person is perceived as greedy, devoid of willpower. From here to deciding that he or she brought it on themselves, and they just have to buck up and do whatever it takes to lose weight, is only a step, and very easily taken. We shall see that things are not quite so cut and dry and that some responsibility actually falls on all of us; and yet the notion that fat people in general are weak-willed and untrustworthy seems deeply ingrained in our psyches. As the French nutritionist Jean Trémolières remarked 'Our society creates the obese, but can't bear them' (Poulain, 2002).

Research tells us that by the time they are six, children select words such as 'lazy', 'dirty', 'stupid', 'ugly', 'liar' and 'cheat' when asked to describe the outline of a fat child, more often than they do for other kinds of silhouette (Staffieri, 1967). This suggests that disgust of obesity is indelibly marked on the brain from a very early age. And a glance through any women's magazine will show how female slimness is presented as a marker of efficiency, success, self-control and sexual magnetism, whereas plumpness betrays laziness and self-indulgent neglect.

However, views of overweight are more positive and don't always carry the same stigma. Indeed studies in countries as culturally diverse as Algeria, Congo, Senegal, Tunisia and the UK, all find that the overweight fail to recognize that their weight is a cause for concern, reducing motivation to control their weight. Indeed plumpness is revered in some traditional societies, but it seems that this is limited to overweight, and appreciation does not extend to the obese.

The social pressure to be thin is so powerful that obese people also drop out of education earlier than their peers, as various studies have shown. The fat have less chance of being accepted by prestigious schools and of making headway in more competitive careers. The same research finds that in both the UK and the US, young women who are overweight earn lower salaries than those with 'normal' figures, or even those suffering from other chronic diseases (Gortmaker et al, 1993). Of course, we cannot dismiss the reverse possibility that such young women may have drifted into overweight precisely because they earn less.

The WHO (2000) has deplored the extent to which this popular rejection is all too often shared by health professionals themselves, including doctors, nurses and even some nutritionists. Their attitude has done untold harm by driving many obese patients away from the consulting-room for good, so as not to endure any more nagging and guilt trips. Not unnaturally in the circumstances, the majority of obese subjects have a poor self-image. They generally regard themselves as ugly to look at, and expect to be shunned by other people – an assumption too often borne out by experience, sad to say. This holds true above all for young women of the most privileged socioeconomic status, a milieu in which obesity, being rarer, attracts even more disapproval than elsewhere. It is also a frequent complex for young women who have been on the fat side since childhood, leaving them scarred by years of playground bullying and taunts.

Low self-esteem can trigger pathological attitudes towards eating. One syndrome is bulimia, frequent in young women who have repeatedly attempted to diet and find themselves see-sawing between weight loss and weight gain. Bulimics stuff themselves uncontrollably at intervals – usually in the evening or at night – before making themselves vomit. Night bingeing is another, less researched disorder, in which the sufferers consume as much as half of their daily calories after the evening meal. One possible cause is a malfunction of the body clock.

These eating disorders, which are becoming increasingly common, constitute serious illnesses that are desperately hard to cure. We are not yet in a position to say whether they may be the cause or the effect of weight gain. But it is highly likely that the acute psychological pressure brought to bear on fat people to make them lose weight is at least partly to blame. The chorus of voices to get thin, in which the media joins in as loudly as anyone, has a catastrophic effect upon the beleaguered targets of all these efforts to make them change 'for their own good'. Especially when as we shall see, they are having the ideal of thinness stuffed down their throats, in an environment that in all other ways conspires to make them fat, so that the goal is practically impossible to attain.

Men's relationship to their weight is far less fraught than women's, even today (although this distinction is probably on the wane), and men are less susceptible to eating disorders. They are handicapped, instead, by their comparative reluctance to seek the medical attention they need for their hardened arteries. There is a case for saying that such eating disorders are fundamentally cultural in origin, as they do not arise in societies where overweight and obesity are signs of success and eagerly pursued.

Underage victims

Children and adolescents pay heavily for their obesity. They run a gauntlet of risks including social and psychological problems, arterial hypertension and diabetes for starters! And for some, digestive and intestinal disorders, sleep apnoea or orthopaedic complications. There is no evidence, to date, that very chubby young children experience significantly low self-esteem. But all this changes when they reach adolescence and their body image plummets with every extra pound. Cruel or thoughtless comments from family, classmates or teachers may be taken doubly hard at an age when young people are vulnerable in every way. The consequences can often last a lifetime. A major study conducted in the US found that women who had been overweight during adolescence or early adulthood were less frequently married, and commanded lower salaries, than women who had suffered from other physical impediments at the same young age (Gortmaker et al, 1993).

Box 2.5 Smoking won't help

Does smoking keep you thin? Many smokers, especially women, rely on cigarettes to keep in shape. Various research papers have shown that smokers pile on the pounds as soon as they give up, with an average weight gain of 2.8kg for men and 3.8kg for women (Williamson et al, 1991). Heavy smokers – on more than 15 a day – and very young ones tend to react even more strongly. The correlation is so striking that some epidemiologists have allowed themselves to wonder whether the successes scored in the fight against nicotine addiction might not be playing their part, indirectly, in fuelling the current obesity epidemic. So is this an argument for not giving up after all? A study published in August 2005 (Canoy et al) in the respected journal *Obesity Research*, throws cold water on any such hopes. It shows that smokers pack more fat in the abdominal region, where it is most likely to lead to health problems, than non-smokers. If smoking does safeguard against a certain amount of overweight, it can hardly be said, then, to encourage a slim waistline. Whatever the case may be, stopping smoking remains a public health priority in view of the destructive effects of this habit upon the body. To sacrifice one's lungs to please the bathroom scales is clinically speaking a pretty poor bargain.

A high price to pay

Treating the pathologies we have listed, both physical and mental, cannot be other than hugely expensive. And the relentless rise of obesity threatens to be a massive drain upon collective resources, in every society it attacks. Everything must be paid for sooner or later. Aside from the straightforward price of the medical treatment of obesity, there are also related costs that will have to be met by the individual or their family, that is, the sum of the different costs caused by the illness, that in extreme cases include death. To these must be added further indirect spending, such as those arising from the loss of productivity or the absenteeism of obese employees. The so-called SOS Study – Swedish Obese Subjects – found that the frequency of extended sick leave, lasting over six months, was 1.4 and 2.4 times higher among obese men and women, respectively, than the average sick leave taken by the Swedish population as a whole (Narbro et al, 1999).

It is not easy to put a price tag on this, but some economists have had a go. According to their cautious estimates, these costs would represent something between 2 and 7 per cent of total health spending (WHO, 2000). Obesity is inescapably confirming itself as one of the biggest drains upon expenditure within national health budgets.

In the US, the total bill of obesity is calculated to fall between $100 billion and $120 billion. But the true costs to society are liable to be higher than what these various studies have predicted, given that the research has not, for the most part, factored in every one of the diseases linked to obesity. It often ignores, too, the expenses due to overweight (BMI range 25–30) as opposed to obesity, even though the number of people who are perilously 'stout' is around three or four times greater than the number of those who are technically obese. Add in the costs represented by the overweight, and the total bill can only rise considerably. On top of that, individual expenditures on losing weight must be taken into account, for these are undoubtedly some of the heaviest indirect costs incurred. Due to the scale of the epidemic, the social costs of what was initially a medical problem must be expected to soar.

Even preventing obesity is costly – in the light of soaring obesity, the UK government has earmarked GB£372 million for a national strategy to try and achieve this, including £75 million for a marketing campaign targeting parents.

...continued... sedentary public transport is more rare... than... structure... ... ...walk not an hour... ...would need to reset... ...walk not an hour... ...

...

Hunt over distant areas of our town... it seems as... rather, our societies geared themselves to do as system as much corporal exertion as possible. The traditional culprit, of course, is television, and the accusation is justified. It has been established that the more we watch, the greater the risk of girth expansion within just a few years. By and large, children of all social backgrounds who frequently watch TV and play video games tend to be plumper, with higher cholesterol levels, than their more active peers. Unfortunately, watching TV is becoming the preferred leisure activity for children as it has for adults. Already in 1994, the average British citizen was spending more than 26 hours a week in front of the box, as compared with 13 hours during the 1960s (Office of Population Censuses and Surveys, 1994). Indeed, it has become a cliché to say that the typical American child spends more time hunched in front of a screen than at school. These kids are tireless at surfing the Web, networking and gaming; they might also read or phone their friends. While their families walk anywhere...

The average number of vehicles per household has risen considerably over the past few decades. It is not unusual for a family to own two or even three cars. Little wonder then that many journeys are undertaken rather than on foot or by bike, no matter how short the distance involved. Many people drive to the corner shop just to pick up some bread.

In the UK, in 1992, children under the age of 14 were already walking 20 per cent less than they did in 1985 (DiGuiseppi et al, 1998). The distances they cycled had likewise declined by 26 per cent, while they were covering 40 per cent more miles in the back seat of the car. The

Chapter 3
Revolution on our Plates

On the one hand we have runaway obesity keeping doctors awake at night; on the other, wary consumers who've been burned so often they no longer know what they should be eating: we are in a proper stew about our food. But in many countries, this crisis goes hand in hand with a less noticeable phenomenon, still in its early stages. For it is our very way of eating that has profoundly changed, and this change has come about almost unnoticed.

The underlying reason for the new eating habits is that in the US, UK, Germany and similar countries, a nine-to-five working day has become the norm. Employees must therefore go to the staff canteen for their midday meal – if they're lucky enough to have one on the premises – and their children eat lunch at school. Others make do with a sandwich or a takeaway on the hoof.

This form of 'eating out' has boomed over the last few decades. It has grown most spectacularly in the US, where the proportion of the family food budget it gobbles up leapt by 50 per cent between 1980 and 2000; eating out now accounts for half of all money spent on food, which adds up to an annual average of $1400 per head. France, by comparison, appears to be lagging in this area. Two-thirds of French adults – women in particular – still claim to eat lunch at home (Guilbert and Perrin-Escalon, 2004). Only one in five of them lunches regularly, that is, more than twice a week, at the workplace. But the trend is on the increase, and there is reason to believe that within a few years the French will have caught up with the Americans, at least in the major cities.

Now, eating in the office cafeteria is not quite the same as eating at home. When it comes to catering for large numbers, there are certain imperatives that must be obeyed, such as keeping costs to a minimum, as the meals have to be sold cheaply. So the menu is unavoidably less imaginative than it might be at home, whatever the company's goodwill, and it is bound to be poorer in micronutrients. The helpings tend to be larger to compensate, and richer in fats – an economical way of improving the dour taste of industrially pre-packed ingredients. Fresh fruit and vegetables, which are inevitably pricier and don't keep for long, often make only a cameo appearance.

School dinners are not much better, on the whole. Their quality may vary in line with each institution's fees, but price remains the overriding criterion for the selection of products. In 2000, the French food watchdog AFSSA published a hard-hitting report on school dinners, which condemned 'the nutritional imbalance of meals served in school cafeterias, from nursery level to secondary level'. More precisely, 'in terms of nutrients, the proportion of lipids is most often described as excessive, iron and calcium are generally deficient, and protein content varies from study to study. In terms of food type, dairy products, fruit and vegetables tend to be under-represented'. Could do better, then! Too much fat, not enough iron, calcium, dairy products, fruits or vegetables. In a separate study, nutritionists found that in some school canteens, fish fingers only contained 50 per cent fish. The rest of the filling was, well, filling – cheaper but nutritionally worthless. Here too, the blandness was masked by artificially jazzed-up sauces with lashings of salt, both to help preserve the product and to make it taste of something. Too bad for our arteries and those of our children.

French schools are not the only sinners in this respect. The UK's Channel 4 broadcast a series of programmes by celebrity chef Jamie Oliver arguing for radical improvement of school meals, which he was happy to describe as 'rubbish' in their present state. This verdict was no doubt a sensationalist one, perhaps designed to raise Jamie's already vertiginous profile, as well as genuinely enlightening the public, and yet the enthusiastic response to his campaign proved that he had tapped a deep well of unease around this issue. When Jamie Oliver collected 270,000 signatures for his nationwide petition, Tony Blair's government felt obliged to act. From September 2006, British school kitchens were told to comply with a list of expert recommendations. Similar measures had already been taken in Germany, where the Consumer, Food and Agriculture Ministry issued a set of stringent regulations in May 2005. For the first time, all the school cafeterias in the country were instructed to serve more vegetarian dishes, to offer fresh fruit at least two to three times a week, and to cut back on high-fat foods and sugary desserts.

But when it comes to fast-food outlets, eat-in or takeaway, the picture is even grimmer. Whether it's a multinational chain or a burger van on the corner, the same simple formula applies: instant snacks, all day long, with an appeal for everybody. This usually means fried food. The customer gets hooked on tastes and textures different to those eaten at home, as these products tend to be fatter and sweeter.

As a test case, let's analyse the basic fast-food meal composed of a regular burger or six chicken nuggets, a 25cl cola and an iced dessert. This provides between 900 and 1200 Calories (kcal), depending on the size of the fries. If we upgrade to a giant cheeseburger and double fries, large soda and milkshake to end, we can top 1600 Calories (kcal), 40 per cent of which come from lipids, most of these as animal fat. The rest of the calories are provided by carbohydrates, but 50 per cent of these consist of simple sugars, contained in the drink and the dessert. Concentrated energy, and little else: each gram of fast food supplies on average 50 per cent more calories than a gram of ordinary food prepared in an American home (Prentice and Jebb, 2003).

McWorld

Fast food in one form or another has flourished all over the world for centuries. But it used to be made fresh from traditional ingredients, like the spicy broths one can still enjoy from Asian street stalls. Today, regrettably, fast food is becoming standardized under the aegis of multinational corporations, whose advertising muscle also enables them to ram their products into more and more countries. The case of the US, which in 2001 could boast more than 13,000 McDonald's outlets, more than 8000 Burger Kings and at least 7000 Pizza Huts, is the gold standard in this respect. From 1970 to 1995, the number of fast-food meals devoured in the US increased fourfold, meaning that currently one in five Americans goes to one of the countless fast-food terminals that litter the country at least once a day.

In the UK, the number of fast-food outlets doubled between 1984 and 1993, while the number of other restaurants and cafés remained stable over the same period. The total of McDonald's restaurants more than quadrupled in Europe in a mere ten years, from 1342 in 1991 to 5794 in 2001. In Asia and the Pacific the company did even better, jumping from

1458 to 6775 outlets. And if Kentucky Fried Chicken and Pizza Hut have been stagnating somewhat in the US, hovering around the 13,000 mark between 1992 and 2001, they are thriving in the rest of the world, where their signs now hang above more than 10,688 doors, up from 5520 ten years ago.

A 2004 survey of Asian adult consumers with internet access found that overall, nearly a third of those questioned treated themselves to fast food at least once a week, almost as many as in the US (where the proportion is 35 per cent).[1] However, Indians and the Chinese leave Americans standing, as 37 per cent and 41 per cent, respectively, consume fast food at least once a week. The record belongs to the citizens of Hong Kong, at 61 per cent! Admittedly this survey made no distinction between fast-food chains and traditional small stalls. Only continental Europe, in this panorama, seems to be holding firm: a paltry one in ten of its inhabitants go for fast food at least once a week.

McDonald's using hospitals as a venue to sell and market their products has come under much criticism. McDonalds has franchises in about 30 hospitals in the US, including children's hospitals in Los Angeles and Philadelphia, and in about seven hospitals in the UK, although many more are situated just outside the hospital gates, strategically placed for visitors and staff. The concern is obvious – hospitals are supposed to be icons for promoting health, and selling fast food on their premises is contradictory to this. Professor Tim Lang of City University, London put this succinctly 'It is frankly pathetic that the public health world does not see the connection between allowing a brand that is famous for selling fatty, sugary foods and drinks on its own territory' (Sweet, 2008).

More insidious are the weekly visits by McDonald's staff to some children's wards in the UK, bearing gifts of toys, balloons and happy meal vouchers. All in the name of entertainment, according to hospital staff. What an effective marketing strategy – for children to associate McDonalds with getting well! The UK government does not object to this practice, even though it is committed to spending billions on preventing obesity. It is considering going even further and allowing McDonalds (and other fast-food companies) to sponsor NHS hospital wards that will allow them to foster some 'brand awareness'.

It has not been scientifically proven, beyond all possible doubt, that eating fast foods is directly responsible for the obesity epidemic. The fact remains that obesity prevalence, in every industrialized country, has gone up in parallel with the increase in family outings to McDonald's and co. A coincidence that is troubling, to say the least.

No more peeling potatoes

Can evening meals undo the damage wrought by unhealthy lunches? Probably not. Because they have changed as well. To the satisfaction of some and the regret of others, the old-fashioned family division of labour – husband at work, wife in the kitchen – has had its day. Women have entered the workforce en masse, and long hours spent at the office and travelling leave them no time for preparing fancy meals. Forget about slaving over vegetables to go into a hearty soup, goodbye complicated casseroles slow-cooking in the oven. We are in such a hurry that our suppers are increasingly built around ready meals and convenience foods. Frozen quiches or moussakas whose ingredients we don't inspect, parboiled vegetables in a sachet to be drowned in some equally mysterious sauce, deliciously sweet and creamy puddings…

A pan EU study of attitudes to food reported that lack of time was a major influence on food choice, due to such irregular working hours or a busy lifestyle (Gibney et al, 1997). So the quicker, the better it seems. Too bad if we can't be completely sure of what it is we're eating.

Our social approach to the evening meal has also changed. Once more the US is lighting the way, for it is there that the tradition of sitting down as a family around the table first began to erode. Everyone ransacks the fridge for something quick and easy to snack on. Needless to say, the potato peeler is a stranger to such kitchens, and there's no nonsense about rolling one's own pizza dough. The most one might do is microwave a ready meal, to be washed down with a soft drink. Thus the principle of a structured meal divided into courses – starter, main course, cheese and/or pudding – is going the way of the dodo; the temptation to simply polish off the left-over ice-cream is just too strong.

Unlike in the UK, things in France aren't quite so bad as yet. The sacrosanct model of the sit-down meal *en famille* is proving robust, and indeed this could be a clue to the (comparative) fitness of adults in this country. Solitary nibbling at all hours is endemic, of course. But the French do maintain – who knows for how much longer – the ritual of shared food, the concept of a sequence of dishes of which everyone around the table partakes, at set times of the day. Nine-tenths of the French population affirm that they sit down each day to three proper meals, breakfast, lunch and supper (Guilbert and Perrin-Escalon, 2004). And one person out of two wisely confines him or herself to those three meals. One in three, mostly women, treat themselves to some light refreshment at mid-morning, or around teatime. And only one person in ten eats five times a day. Overall, women are consistently more likely than men to snack between meals.

Are we saying that the French have got it right? Well, not really. There is a growing tendency to simplify the meals they eat. One-third of adults content themselves with a two-course lunch (steak-*frites* followed by a dessert, for instance). The proportion goes up to almost 40 per cent when it comes to the evening meal. This is not necessarily a bad thing, provided the course that is skipped is the one involving cream cakes. It's more of a problem when the piece of fruit that wraps up a traditional meal is left out. Or if, instead of the traditional 'cheese and fruit', one drops the fruit and keeps the cheese: this is scarcely the stuff of a balanced diet. Cutting out the first course, too, usually means sacrificing a portion of vegetables. In any case, it is never a good idea to reduce the variety of the foods we eat.

A further issue of concern is the amount of time people spend in front of the TV. It seems to increase with every survey (two hours a day is the average for French men and women over the age of 12, according to the health and nutrition survey of 2002). The habit of watching television while eating seems to have become firmly entrenched. More than one in five (especially young children) have breakfast with one eye on the box, almost half (especially young adults) watch it at lunchtime, and more than half (especially older people) have it on during the evening meal. Since 1996, watching TV at breakfast and lunch has notably increased. Research shows that the longer someone spends in front of the box, the larger his waist circumference. Of course, there is no evidence that watching telly while we eat is fattening in itself. But it cannot be denied that there is a strong link between TV-centred lifestyles and obesity.

From one revolution to the next

These shifts, common to all developed countries, are profound and lasting.

They are dwarfed, however, by the upheavals that have taken place in developing countries. Over the course of a few short years, without perhaps being fully conscious of the scale of these changes, such countries have completely transformed not only their diets but also their ways of producing, distributing and consuming food. The enormity of the changes are such that scholars have begun to refer to a 'nutritional transition', analogous to the demographic transition that occurred in developing countries a few years ago, and is still under way in some of them.

This is not the first time, of course, that humanity has undergone a major overhaul of its eating habits. There have been several such historic turning points since the days when early tribes subsisted by hunting and gathering. That way of life yielded a relatively balanced diet, and food was positively abundant in certain regions. And then hunter-gatherers invented agriculture, as a way of producing their own food in a controlled and convenient fashion. For thousands of years after that, people relied on cereals and beans/pulses to supply their basic needs in terms of energy and protein, with the addition of livestock in some places, or the predominance of root vegetables and tubers in others. The adaptation took its toll: the earliest farmers were not as tall as their physically active ancestors, no doubt because cereals are rich in phytates, which trap the minerals necessary for growth such as calcium, zinc and iron. Absorbing fewer minerals, and without the compensation of regular access to meat, people tended to be shorter. Even so, the adoption of agriculture led to the first demographic boom. The population grew, until a balance was reached between the amounts of food that could be produced and the number of mouths to feed. It was a precarious balance, constantly rocked by epidemics, wars and other disasters, so that we find a pattern of shortages and famines alternating with eras of relative plenty. Excluding times of war, the last famine to grip the West took place in 19th-century Ireland, where it provoked a mass emigration to the US.

The Western world now embarked upon the Industrial Revolution. First, yields rocketed as a result of technical breakthroughs in farming and the mechanization of agricultural chores, making many labourers redundant and driving them to the cities – a move that in turn fuelled greater industrialization, in one country after another. The expansion of cities put new pressures on the food supply, leading to a still greater mechanization of agriculture: the upshot was that productivity soared.

This is not to say that everyone was eating their fill, far from it. The novels of the 19th-century French writer, Emile Zola, are there to remind us, if need be, how painfully this period was felt among certain sectors of society. All the same, the population's food needs were being more widely met than ever before, thanks to the increased production of calories and basic nutrients.

This achievement was helped by the fact that most urban occupations were less physically strenuous than rural labour, so that total energy needs diminished. Between rising food production on the one hand, and falling physical activity on the other, a perfectly adequate balance was eventually reached. This was the age of falling hunger, when industrialized societies were able to nourish all of their citizens to a reasonable standard.

Unfortunately the story does not end there. As today we are becoming aware of the limits of this mass production and consumption system, whose extravagances have not been properly managed. The ageing of the population has unleashed a spate of chronic diseases that were formerly quite rare, while at the same time, the overload of calories resulting from ever more unbalanced diets is triggering the onset of such diseases at increasingly young ages: in adolescence, as we saw in the preceding chapter, and often earlier.

A change of pace

In Western countries, industrialization established itself relatively slowly, over more than a century. Societies had enough time to adapt to all or at least part of the changes that ultimately contributed to raising the standard of living across the board. Contemporary developing societies have no such luck. The globalization of trade is forcing them to make the transition at breakneck speed. In the space of one or two generations, they have tipped from a situation in which malnutrition and undernourishment were the most urgent problems, into one in which obesity and its associated diseases have become the main issue of concern.

The first warnings came from some tiny Pacific islands. In these comparatively isolated, untouched spots, the supermarket culture allied to a radically unaccustomed lifestyle suddenly took over, with immediate repercussions including an explosion of obesity and diabetes. The phenomenon was too remote, however, to attract much notice. Next came the turn of Latin America. Chile, Uruguay, Paraguay and Argentina had embarked upon their nutritional transition a little more than 30 years ago, during the 1960s and 1970s. Brazil and Mexico followed close behind; today, the proportion of obese Mexicans almost rivals that of their North American neighbours. The islands of the Caribbean, too, were soon receiving their share of low-cost products imported from all over, in an onslaught that wiped out local subsistence cultures. The dietary balance, such as it was in these precarious islands, soon shifted, especially as the nutritional transition was imposed upon a society that was far less economically developed than Mexico or Brazil had been in the same situation.

At the time, few doctors saw any cause for alarm. Health and nutrition programmes were overwhelmingly focused upon the elimination of undernourishment, by hook or by crook. Neither overweight nor obesity was monitored. The very notion of type 2 diabetes existing in previously underfed groups was inconceivable. And yet more and more cases were turning up in doctors' consulting rooms. Simultaneously, cardiologists were seeing a rise in heart conditions that formerly did not exist in those parts of the world. But, segregated into their various specialisms, the professionals failed to compare notes and the global epidemic continued to spread unnoticed. It was only when the mounting number of cases became glaring that the links were eventually made. Doctors abruptly realized that they were faced with a serious and ubiquitous problem. But how were they to convince the politicians?

In the mid-1990s, on the occasion of a conference held in Alexandria to examine the issue of households at 'nutritional risk' in Egypt, one paper called for maximum subsidies to be applied to staples such as flour, oil-seed, sugar and other high-energy products in order to pre-empt food riots by making sure that everyone, no matter how poor, could feed themselves for a reasonable price. At the same time, nutritionists were observing with dismay the headlong rise of obesity and the pressure on crowded diabetes clinics in hospitals. This exemplifies the hopeless discrepancy between food policies that remain obsessed with ensuring a supply of calories for all, and the real emergency, consisting of the worldwide increase of overweight. As WHO warned in 2000:

With the improvement in socio-economic status and increasing changes

due to rapid urbanization, the prevalence of obesity among some groups of
black women has risen markedly to levels exceeding those in populations
in industrialized countries. In fact, approximately 44% of African women
living in the Cape peninsula were estimated to be obese in 1990.

How can the rate of obesity be so high in poor countries? To begin with,
it's much cheaper than it used to be to fill up on calories, sugary and fatty
foods. The productivity of the great agricultural powers such as the US or
the EU has increased so much that they are staggering under mountains of
excess sugar, grain, oil and animal fats. The answer has been to export these
surpluses, flooding local markets all around the world. If there has been so
little resistance, it is partly because once a peasant emerges from extreme
poverty, and especially after migrating from the countryside to the city, he
or she is typically reluctant to eat the same things as before. In Brazil, many
rural communities subsist mainly on beans and manioc. If their incomes
happen to rise, they will add rice to their diet, since it is easier to prepare,
and also tastes better. As soon as they can afford it they will throw in some
bacon, which is not as expensive as it is satisfying. And once they become
fully urbanized they will start feasting on bread, wheat, meat and so on,
just like the rest of us.

A country like Mexico, however, has already moved beyond that stage.
Research has found that overweight and obesity are as frequent in rural
Mexico as in the cities, just as they are in Europe and other developed
countries. Farmers drive cars and work the land from the comfort of their
tractors, their homes are fully equipped with mod cons, and they buy their
food at the mini-supermarket on the corner. Where else? In Mexico, as in
most of Western Europe, the days when farmers consumed their own
produce are long gone. Instead, they might grow maize, say, for one food
conglomerate – and feed the family on ready meals packaged by another.

Fat at last!

By now, as a result of all these changes, the world presents a patchwork of
very different patterns. There are still some marginal societies that cling to
the hunter-gatherer lifestyle, as in Amazonia or Papua New Guinea.
Elsewhere we find communities who are still at risk of hunger or starvation,
whether landless peasants, or people with access to land but living in places

where the population is too dense for this land to feed everybody adequately. Examples would be Rwanda or Burundi, where although the peasants are so technically proficient that they can grow food on mountaintops, they are still struggling to provide for a population almost as dense as that of Belgium. Other parts of Africa and Asia are blighted by war, and thus are still vulnerable to famines. Lastly, the newly industrialized regions such as the oil kingdoms, North Africa, Latin America, Thailand and the other 'Asian tigers', and of course China: all these have conquered basic hunger and are stepping confidently into the world of consumer abundance.

As these countries become industrialized, and their cities grow into megalopoli, globalization incites the newly urbanized masses to adopt a Western lifestyle, complete with a taste for processed food. The majority of Chinese now own a television, an object that few had ever laid eyes on 20 years ago. They travel to work on public transport, which has expanded to carry them, although shortly they will abandon it for private cars. There are clear benefits to be gained from this evolution. Everyday life is more comfortable, food is more abundant and varied. Supermarkets and superstores have mushroomed, offering thousands of attractive products to a population that lived, not so long ago, with almost nothing. But there are two sides to the coin. Quite quickly, the new diet proves unsuited to the new sedentarism: people begin to consume more meat, more fatty foods, more sugary drinks, such as Coca-Cola. If ten years ago, 10 to 20 per cent of the calories in the typical Chinese diet were obtained from lipids, these substances now constitute 30 per cent of the calorie intake of prosperous city-dwellers. Conversely, the intake of slow-release starches and of fibres has plummeted. Such abrupt shifts are not immediately registered by the population at large. But they are gradually beginning to be acknowledged as a significant public health problem.

One obstacle to recognizing obesity as problem is the fact that in many such societies, obesity may still – if only for the time being – be actually admired. In historically deprived countries where there was, or still is, a shortage of food, plumpness is perceived as a mark of wealth and social status, and indeed of good health. As the WHO (2000) reminds us, 'Fat women are often viewed as attractive in Africa.' It goes on to observe that 'in Puerto Rican communities, weight gain after marriage is seen as showing that the husband is a good provider and that the woman is a good wife, cook and mother. Weight loss is socially discouraged.'

Box 3.1 Can a foetus be hard-wired to get fat?

Why do some people put on weight more easily than others, and develop more chronic diseases in adult life than others, all else being equal? Seeking an answer to this question, a team of British doctors under Professor David Barker (1992) discovered that many obese subjects had not been adequately nourished or cared for while in their mother's womb, and then during their first year of life. The researchers thus proposed the theory that adverse prenatal and post-natal conditions might cause the metabolism to adapt to conserving fats, so as to counteract the initial shortfall. The adaptation being permanent, however, it proves detrimental if ever food becomes more abundant later on.

Since then, many studies have found evidence to support the theory of 'fœtal programming'. It would help to explain, for instance, why the 'obesogenic'[2] effects of the nutritional transition are so pronounced in countries whose adult populations, currently with access to calorie-rich foods, would have been deprived *in utero* and during infancy. Their bodies remain indelibly marked by this early deprivation, irrevocably programmed to hoard reserves of fat.

The fat and the thin

These upheavals first affected a number of newly emergent countries with ample resources, whose economies were fairly well integrated into international trade networks: such as South Korea, Brazil, Mexico, or more recently, China. These countries had successfully dealt with the worst manifestations of poverty-related malnutrition and infectious diseases, even if such advances did not benefit all social classes to the same degree. Little by little, however, the changes we speak of spread to lower-income countries, and became implanted in the urban centres, at least, in the poorest nations of all. Here, the consequence has been that chronic diseases, such as diabetes or hypertension, have not so much displaced as compounded the old problems of undernourishment and infectious disease.

In very poor nations such as Bangladesh, Burkina Faso and Madagascar, rural populations continue to be drastically underweight, victims of seasonal shortages and crop failures. We should never forget that some 850 million people around the world still don't have enough to eat. But there is an astounding contrast between the country and the cities, where overweight predominates. Ten years ago in Ouagadougou, capital of Burkina Faso, one in five women were too plump. Today the figure is one in three – and yet this is one of the poorest countries in the world. The same asymmetry applies in China, where 130 million country-dwellers suffer from chronic undernourishment while their urban compatriots are becoming obese. There certainly seems to be a link between urbanization and obesity in the developing world.

> ## Box 3.2 Food aid: Between the devil and the deep blue sea
>
> In our efforts to combat hunger around the world, are we perhaps over-feeding an entire generation, turning it into the world record-holder for collective girth? This was the chief question posed by the London conference held in February 2005 by the International Obesity Task Force, designed to address the issue of dietary needs. Both the WHO and FAO recognize that young children have been systematically over-pumped in terms of energy for the last 20 years. Surely this must have contributed to aggravating the obesity problem in some countries? The same question arises with respect to programmes in place to deliver food supplements to young children, and which privilege energy content over nutritional balance. According to some Chilean scientists, such interventions have boosted the increased obesity observed in that country. The targeted children were suffering from growth retardation, the most common form taken by child malnutrition in the developing world. But the excessive energy provided by the programmes only bulked up the recipients, increasing their BMI while failing to correct their impaired growth.

In the urban centres of such countries, it's the better-off who are the first to change their eating habits, because they can afford the goodies on the shelves of the brand-new supermarkets. By the same token, they are the first to become overweight or obese. But, because they are also better educated and influenced by Western ideals of thinness, they are quick to assimilate the importance of eating less and balancing their choice of foods a little more;

besides, they have the money to do so. The phenomenon, however, spreads rapidly to the middle classes, followed by lower socioeconomic groups. People on modest incomes suddenly find a cheap, calorie-packed diet within their grasp and make the most of it as soon as they can. Unfortunately this means sacrificing many elements that are nutritionally more valuable, such as fruit and vegetables, both because they are more expensive and because people are not conscious of their importance. Surveys carried out in Brazil neatly illustrate this trend: they reveal that problems of hypertension and other chronic diseases resulting from obesity are becoming increasingly common in the more deprived urban districts. The foodstuffs consumed by such communities are often wanting in terms of anti-oxidant vitamins and minerals that ward off chronic illnesses. An Argentine sociologist has discovered that since the late 1990s, in her country, the standard diet of the poorest people has dwindled down to a handful of products more notable for 'filling the belly' than for their vitamin content (Fraser, 2005). In order to make ends meet, humble households have to buy the cheapest foods, which also happen to be the fattiest.

What is more, epidemiologists have found that underweight and overweight can coexist within the same family, known as the double burden of malnutrition (Garrett and Ruel, 2003). A child may be visibly malnourished and presenting signs of growth retardation, while one of his parents (usually the mother) is overweight or obese. The latest available data suggests that in a country like Brazil, some 11 per cent of families with a stunted child under five are in this situation. In Egypt, the figure is closer to 50 per cent! How are we to interpret such findings? Is there a shortage of food on the table? Obviously not, since the mother is too fat. At the very least one can assume that the amount of available calories is more than enough to meet the household's energy needs. However, there is a probable difficulty in obtaining foods that are richer in vitamins and micronutrients, such as fruit and vegetables. The mothers are often anaemic and deprived, despite their bulk, of essential substances such as iron, zinc, vitamin A or folic acid (de Souza et al, 2007; Zimmermann et al, 2008). Or perhaps the key to the mystery lies in the mother's lack of knowledge about child nutrition, causing her to feed the same traditional gruel every day, without realizing that children need a more balanced and varied intake if they are to grow, unwittingly reproducing for her own infants the same kind of nutrient-poor baby food that stunted her own growth, 20-odd years ago. And it is safe to assume that later on, older children exposed to their parents' fat-rich, sugar-rich diet will adopt it for themselves and become overweight in no time. For the person who was poorly nourished during infancy has an increased risk of piling on the pounds in later life.

It would be too easy to dismiss this phenomenon as a problem for developing countries alone. The same paradox prevails in the US. During the mid-1990s, the Census Bureau calculated that 11 million Americans were living in a situation of food insecurity, that is, they did not enjoy access to enough food for them to lead a healthy and active life at all times. And to these 11 million citizens at risk from malnutrition were added a further 23 million who, while having enough to eat as a rule, could be regarded as periodically falling into a state of food insecurity (Eisinger, 1998). One of the wealthiest countries in the world was failing to guarantee the sustenance of its people, both adults and children: over 4 million youngsters under 12 had gone hungry, and nearly 10 million more had been at risk of going hungry for a period of at least one month during the year preceding the survey. For its part, in 2002 the US Agriculture Department's economic research service worked out that 11 per cent of American households had experienced a state of food insecurity at least once.

Box 3.3 Another economic burden for developing countries

There have been few attempts to compute a credible and comprehensive price tag for the treatment of obesity-related diseases. And yet international agencies have already expressed their forebodings, given the speed with which these diseases progress. In many developing countries, existing medical resources are already drained by the needs of urban populations and of the increasingly well-heeled middle classes; how are they to cope with the extra demands, when national revenues are so much lower than in industrialized countries? For these diseases are proportionately more costly to treat in the developing world. This is because the necessary equipment and drugs have to be imported, and specialized medical staff has to be trained. In India, a course of treatment for hypertension can cost the equivalent of a whole year's wages. The very slight advantages inherent in an overfed population are therefore set to be far outweighed by the financial disaster of an obese society.

Should the nutritional transition in the developing world be seen as a replay, in speeded-up form, of a process that had unfolded a few decades earlier in the US and other developed countries? Possibly so. And yet the surge in childhood obesity, which has been rising at much the same alarming rate everywhere, in rich countries as in poor, suggests that other forces may be at work. Why are children in Poland, in England, in Egypt, in China and in the US all at once falling prey to very similar problems of overweight and obesity, as well as hypertension and diabetes? Powerful worldwide forces, which impact as much upon our diets as upon our lifestyles, seem to be calling the shots nowadays. It is time to look at these factors more closely.

Notes

1   www.nielsen-online.com
2   Obesogenic is defined as 'promoting obesity'. Concept introduced by Boyd Swinburn of Deakin University, Australia.

Chapter 4

Agriculture in the Age of 'More is More'

If you want to be thin, don't eat so much. The complacent refrain is a familiar one, and its solution to obesity seemed so blindingly obvious that for a long time this illness was relegated to the status of a purely medical problem, on a par with alcoholism and tobacco addiction. It was the doctor's job to work on patients and make them control their self-indulgence! It was up to nutritionists to devise the ideal diet, which had only to be followed to deliver fitness for all! The failure of this approach, despite the best efforts of a highly competent medical profession, may seem baffling. And yet it was largely to be expected. This is because the method was founded on the assumption that a person grows obese due to a series of bad choices, and it would be enough to point out the error of their ways for them to switch perfectly naturally to the right path. This was a tragic misunderstanding: witness the fact that obesity is mounting fastest among the poorest, least-educated sectors of every society. There are clear social and economic factors underlying the trend and the reasoned arguments of nutritionists cannot stand up to this more obstinate fact.

The preceding chapter sought to chart the profound modifications to our eating habits, both in economically wealthy and poor countries. But it would be naive to suppose that these changes are a function of the consumer's free will. To believe that we eat what we please, when we please and how we please is to overlook that fact all food must first, before it is eaten, be produced, processed, distributed, sold and finally prepared. And that the consumer is merely the last link in an immensely long chain with countless economic, social and cultural implications.

For us to reach the point of unwrapping our pizza (Napolitano, let's say) and sliding it into the oven, first of all the raw materials had to be produced. Wheat, tomatoes, olives, a few anchovies, the milk for the mozzarella… All these were tended and harvested by a farmer (except for the anchovies, caught by a fisherman). The tomatoes were taken away in a lorry to be puréed in a food processing plant. A machine somewhere else mixed the pizza dough (we are not speaking of delicious hand-made pizzas, which are fast losing the market share). The various ingredients were no doubt finally assembled into pizza at another factory. Then the items had to be packaged and freighted by road to the supermarket, where they were swiftly arrayed on the shelves by hand. From there our pizza was transferred to its almost final destination: the hurried shopper's trolley. Of course, the above is only a condensed summary. On a less concrete level, we must remember that 'chefs' and marketing experts originally designed the recipe, and PR teams chose an appetizing name for it, or for the brand behind it. Someone else checked that it met with the latest customer tastes, someone else took care of advertising, and someone else negotiated the deal with the retailers for mouth-watering promotions and discounts.

The point is clear: there are a staggering number of activities revolving around the issue of food. In France alone, the food industry had a turnover in 2003 of €136 billion, and provided 421,000 jobs.[1] The issue of what we eat is rather more than our own private business!

Another feature of the way this system works is that small dairy producers, who peddle their hand-crafted cheeses at the local market, are participants just as much as the huge multinationals such as Danone that supply the whole world. Of all economic sectors, the food industry is undoubtedly the one involving the greatest diversity, if not complexity. What does a small cassava producer in central Africa have in common with a mid-western American farmer, riding a huge automatic thresher across thousands of acres of maize, constantly abreast of prices on the Chicago Stock Exchange? Nothing at all, beyond the fact that both of them live off working the land. And yet both form part of the world food system.

Towards an agriculture without farmers?

This agricultural economy evolved over the whole planet in four stages, to put it simply. The first is the 'agrarian' stage, in which growers consume only what they produce. This system is nominally self-sufficient, although yields are barely adequate to feed a family. Two billion extremely poor peasants still practice it today, in Burkina Faso, Bolivia, Bangladesh and elsewhere. In the plains of the Sahel, the men grow sorghum that the women crush by hand, before hulling and cooking it. There is no processed food to be had. It should be noted that these rural peoples seldom have trouble with obesity. Their most urgent concern is to obtain sufficient calories.

With the onset of economic development, society reaches the stage known as 'artisanal', in which a certain division of labour becomes established. The countryside grows enough food to service the expanding towns. And as these towns evolve, their inhabitants become accustomed to a certain amount of 'eating out'. This is the case of Morocco, for example. And yet a significant proportion of that country's population continues to subsist entirely on its own production.

The tendency becomes more pronounced with the emergence of the 'industrial' stage, currently gaining ground all over the world. Specialization is reinforced; farmers no longer sell their produce directly, but through a string of middlemen. As just one part of the chain that leads to the final product, packaged and ready to be consumed, the role of agricultural labour properly speaking is smaller than that of industrial processing. Food products become enriched by a raft of services, for example advertising, market research and graphic design, which enhance their added value.

By the time we reach the final phase, known as 'agro-tertiary', farm work itself represents a quasi-negligible amount (around 20 per cent) of a food product's final value. It has shrunk in comparison to the share that corresponds to industrial transformation (approximately 35 per cent), and this in turn has been dwarfed by the ensemble of services that give supplementary, intangible 'value' to the various products. This is the world of stringy cheeses sticks, conceived specially to appeal to children, or of the partnerships between entertainment empires and food companies in which the celebrities owned by the first will be used to promote the calories of the second, and so on. The percentages above are merely averages, of course. The figures vary in accordance with the type of food. In the US, the proportion of the retail price that went to producers of beef, eggs and chicken was as much as 50 or 60 per cent in 1998, whereas vegetable growers only clawed back 5 per cent (this means that of every 10 dollars' worth of supermarket vegetables, a miserly 50 cents would go into the farmer's pocket).

In this last stage of economic development, eating away from home turns into a major habit; accounting for almost half of all food-related expenses. To date, only the US can be said to have attained the fourth level of development. And whether or not by coincidence, it is also the country in which the problems of obesity were felt sooner than anywhere else. We shall see further on what part the food industry and its satellite services may have played in the emergence and propagation of the pandemic. Western Europe stands, for the moment, halfway between the industrial stage and the agro-tertiary stage. But all the evidence suggests that it is going down the same road as the US and will eventually replicate the American model.

Produce more!

Whether they be primitively agrarian, in the industrial mainstream, or agro-tertiary already, for a long time all of these food-producing systems were based upon a common goal: to produce more calories. And to do so preferably in the most efficient, cheap manner. When the National Institute for Agronomic Research (INRA) was created in France in 1946, rationing was still in place, while at the same time an ebullient birth rate was multiplying the number of mouths to feed. The aim of the institute, like that of agriculture in general, was crystal clear: to mobilize science and technology in the service of giving everybody enough to eat. The same pattern was seen in the UK and other Western European countries. Agronomic research conscientiously delivered on this mission, giving priority to high-yield grain species and the most intensely productive crops and livestock. The size of holdings was increased and methods of production were mechanized.

In the US, similar measures had been implemented long before. When the US Department of Agriculture was set up in 1862, its main brief was to guarantee a plentiful and salubrious food supply for the population and encourage people to adopt a more abundant and varied diet, so that even the poorest might have access to decent food. American industrial efficiency was put to work in the service of enhancing agricultural performance by supplying machinery, fertilizers, pesticides and other aids.

These policies surpassed all expectations in France. Between 1961 and 2002 the yield of wheat was tripled, from 2.5 tons to 7.5 tons per hectare.

Already by the early 1970s, France had become self-sufficient in food and was beginning to export massive quantities of its agricultural surplus. In other words, both in the US and Western Europe, an age of penury gave way to one of overproduction. Prices fell of course, which was greatly to the advantage of consumers. But peasant communities found themselves obliged to increase production still more in order to compete, which only led to further price drops.

The perverse effects of farm subsidies

One element that played a key role in the spectacular surge of agricultural output was subsidies, which enabled crop or livestock farmers to sell their products for more than the market price. Thus assured of a stable, guaranteed income, farmers were free to invest in productivity without bothering too much about the commercial viability. The subsidy system enshrined in the European Common Agricultural Policy (CAP) was instrumental in boosting output, since growers were sheltered from the fluctuations of the market. Ultimately, however, it fostered the build-up of mountains and lakes of products that nobody knew what to do with. This process came to a head in 1983–1984, with the milk quotas crisis. Europe was drowning in so much milk that by 1983, the surplus was absorbing 30 per cent of the European Agricultural Guidance and Guarantee Fund (EAGGF), and placing the future of the CAP in jeopardy. One million unsold tons of dried milk had piled up, plus 850,000 tons of butter, enough for more sandwiches than anyone cared to imagine. It became necessary to impose limits on the producers, in the form of quotas. But this did not solve the problem altogether. Because milk continued to be so cheap that producers, up to their necks in debt to finance the business, could only survive if they accumulated enough quotas. It was a vicious circle, in which overproduction entailed further excessive production just to break even. The logical consequences soon kicked in, and the smaller outfits went broke. The number of dairy producers fell by two-thirds between 1985 and 2003.

Today, neither Europe nor the US have really found a way to extricate themselves from the perverse logic of subsidies and surpluses. Excess production is sold for a pittance to the food industry giants, which recycle it into frozen meals or tinned/processed food. In the EU, some 500,000 tons of butter – a third of EU production – are thus acquired at knock-down prices by the industry and used to make bakery and dessert lines, at a cost of €500 million (Schäffer-Elinder, 2005). To shift these mind-boggling quantities of cheaply produced food, there is only one solution: the agro-industrial system must do its level best to pump up consumer demand, wheedling shoppers into buying – and hence eating – more stuff. These shoppers are not only European or American; the pressure to consume also reaches the developing world, to which some surpluses are exported. Liselotte Schäffer-Elinder, an Associate Professor at the Swedish National Institute for Public Health, condemned the whole mechanism in an article for the *British Medical Journal* in December 2005: 'Phasing out of market support for agricultural producers in developed countries is necessary as a first step in the fight against obesity, poverty, and hunger worldwide.'

A crucial detail in this regard is the fact that fruit and vegetable crops, unlike cereals, meat and dairy, have not been put through this inexorable productivity mill. They have attracted fewer, smaller subsidies. This is because there was never any intention of feeding the increasing number of people in the world with fruit and vegetables. More solid, substantial fare was required, which mostly translated into grains and meat. It's no surprise then that fruit and vegetables are so sparingly represented on our plates – or that their growers are often hard pressed to sell such crops.

Europeans and Americans were merely the first to exemplify a phenomenon that appears to be universal, regardless of region or culture, which is that when people have extra money to spend (which they very quickly did during the boom years from 1950 to 1980), they invariably desire to spend it on meat. They never clamour for more spinach and beans. As household incomes rise, the typical family starts to first buy more grain, then more dairy products, and finally they splash out on meat, which although more expensive is irresistibly attractive to newly solvent populations. We are currently witnessing the same pattern in China: as their spending power increases, the Chinese do not redouble their consumption of soya – they move on to pork instead.

This soaring demand is not, of course, likely to be met by hunting. It requires intensive rearing of pigs, cattle and poultry. Here too, production in developed countries has been equal to demand, at least in terms of quantity. However, from a nutritional point of view, it is too often forgotten that reared meat has little in common with game, above all when it comes to saturated fat content. In light of the striking fact that the obesity epidemic promptly declares itself in all populations that have begun to consume more meat than they used to, it is hard to believe that there is no correlation at all between the two phenomena.

Box 4.1 Tender chicken

During the decade of the 1970s, British doctors officially urged their compatriots to consume more poultry, presented as the low-fat alternative. Intensive poultry farming sprang up all over the country in response to the demand, which rocketed to such an extent that today, we devour 25 times more chicken than we did in 1950 – with consequences that the medical profession had not anticipated. Barnyard chickens and battery chickens are rather different creatures in terms of composition, for the latter carries considerably more fat. This was demonstrated in a study conducted by London's Institute of Brain Chemistry and Human Nutrition. By analysing poultry samples bought in supermarkets, researchers discovered that the nice plump chicken of today contains 100kcal more for every 100g than its ancestors of 30 years ago. In counterpoint, it offers three to eight times less fatty acids in the form of omega 3 (regarded as an important safeguard against certain cardiovascular diseases), replacing these with the not so beneficial omega 6. Cheap chicken may be lovely and tender, but it's not quite the healthy option we once thought.

The worldwide demand for animal products in the developed world has indirectly fostered another damaging effect. In order to feed the growing armies of cattle and pigs, farmers began to plant huge expanses of oilseed crops such as rapeseed, sunflower and soybean. By pressing these grains two products are obtained, vegetable oil and the protein-rich oil cake that is used as the basis for animal feed. The growers rapidly found themselves saddled with vast quantities of vegetable oil, which they had to sell at rock-bottom prices in order to get rid of it. Thus the market was inundated with a tremendously high-energy food source.

The success of the 'green revolutions'

The dizzy increase in yields was not the prerogative of the developed world alone. All over the planet, from Asia to Latin America, production was stepped up. It was a genuinely global effort, spurred on by international agencies whose objective was explicitly one of quantity, given the need to prevent the famines that would otherwise befall an irresistibly spiralling world population.

During the 1960s, international agencies – with the support of private organizations such as the Ford and Rockefeller foundations – poured money into agricultural research in developing countries. The aim was to come up with new high-yielding varieties of cereals (rice, wheat and maize), which, coupled with a range of technologies including irrigation, fertilizers and pesticides, would boost productivity in the fields of the South. These policies, largely modelled on Western practices, came to be known as the Green Revolution. The new agriculture chalked up important successes in Asia, the Middle East, Latin America and North Africa,[2] where harvests of rice or wheat more than doubled. Thanks to this, not only did the dreaded famines not materialize, on the contrary: the available food per head increased by a significant amount between 1960 and the present day. The new techniques modernized farming, helped pull millions of smallholders out of poverty, and made it possible for various countries to grow enough to feed their entire populations, as well as to export the surplus in order to fund further development. However, not everything is coming up roses, and contentious issues remain. Critics of the Green Revolution point out that the new practices rely on machinery and agrochemicals that are manufactured in the North, reinforcing the dependency of poor countries on a handful of multinational companies. They also encourage farmers to get into debt, in a way that may be crippling. Many small farmers proved unable to keep up and lost everything they owned, before moving to swell the population of some urban slum. Finally, some argued that this technologically enhanced agriculture was implanted at the expense of ecosystems, impoverishing the soil and reducing biodiversity, introducing crop strains that were more prone to pests and diseases (traditional varieties are more resistant), and contaminating the ground with all kinds of chemicals (the Chinese were using 300kg of fertilizer per hectare in 1992, according to the FAO!).

Agronomists are currently working on ways to close the gap between the reality in the field and the results obtained under laboratory conditions, while seeking to minimize environmental damage. This research is not confined to crop farming; livestock farming has also spectacularly increased its productivity, due for example to the genetic modification of certain fish, such as Atlantic salmon or the African tilapia, to obtain gains of 45 to 75 per cent.

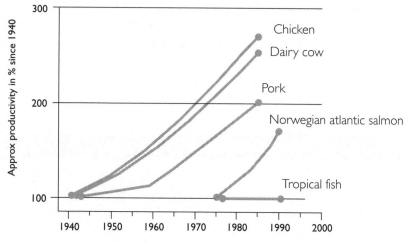

Figure 4.1 Global livestock production

Source: FAO 1996

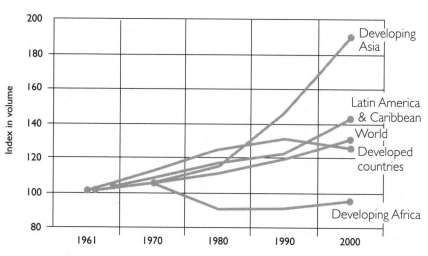

Figure 4.2 Evolution of food production per head

Source: FAO stats

But the fight goes on

And yet we should not jump to the conclusion that the war against hunger has been won. The global obesity epidemic must not distract us from the fact that there are still 850 million undernourished people in the world. Demographers expect the global population to rise from the present figure of 6 billion, to 9 billion by 2050, and most of this growth is projected to take place in urban areas. That means a lot of new mouths to feed – especially in Africa, a continent that experts fear will be unable to provide food security for all of its inhabitants. It is thought that between now and 2030, undernourishment will continue to decline overall, falling to 6 per cent in developing countries as a whole, and yet it will still stand at around 15 per cent in Africa.

Box 4.2 Can the entire world be fed one day?

By producing ever more food, do we stand a chance of eradicating hunger at long last? There is no simple answer to this question, because feeding the planet is about much more than producing food in sufficient quantities. The FAO world report on food and agriculture in 2000 (FAO, 2000) admitted that world production is more than adequate to feed the world's population and yet 'More than 800 million people are still seriously underfed in dietary energy supply terms'. This suggests that even a very sizeable increase in agricultural output would not suffice by itself to deal with the prevalence of undernourishment. Each population must have the resources to pay for its food needs, and this food must be able to reach the hungry. Unfortunately then, we must accept that some countries, many of them in sub-Saharan Africa, will continue to suffer from hunger even while their neighbours have more than they need. The problem at present resides more in the way that markets are structured, than in gross levels of production worldwide. There is also the brutal factor of war, which can suddenly put whole regions at risk of starvation while making it impossible for international agencies to intervene in good time. The FAO has recently acknowledged that armed conflict is on a par with poverty as one of the main causes of the hunger that continues to stalk certain parts of the world. According to the same organization, wars will no doubt prevent Africa from reaching the targets set for 2015 by the Millennium Development Summit.

A billion Chinese meat-eaters?

In view of the mountains of surplus produce generated by US and European agriculture, common sense tells us that the food needs of the planet can easily be met. But nothing can be certain. As an illustration of the problem, let us take the case of India or China. On the face of it, both countries grow sufficient quantities of grain to keep their people in calories. Or do they? It's hard to say. Nobody marches into a store to purchase their calories by the dozen, like eggs. The consumer doesn't think like that. The consumer buys a range of products, not always the same things; they are after particular tastes and textures, wanting to cook a specific dish. As a result of such behaviour, the real demand of any given population is never commensurate with its purely physiological needs. And that is what makes matters of food and diet such complex issues.

As we have seen, demand varies as a function of income. Meat in particular is wildly popular as soon as the family can afford it. This has the knock-on effect of boosting cereal production, since the animals, too, have got to be fed. One animal calorie requires an average of seven vegetable calories. In other words, the animal has to consume seven calories in the form of fodder to produce one calorie in the form of flesh. Intensive livestock farming uses the equivalent of 9kcal of grain to make 1kcal of beef, a proportion that becomes 4/1 for pork and 2/1 for chicken. Therefore, when a country becomes prosperous enough for large parts of its population to be eating meat regularly, the amount of cereals that must be produced rapidly overtakes the level at which there would have been enough calories to go around (around the humans, that is). Alternatively, urban lifestyles usually involve a fall in physical activity levels, and so the growing mass of city-dwellers ought – in theory – to need fewer calories in order to subsist. Which of these two trends will weigh more in the balance? To accurately anticipate and plan for a population's true food needs is a daunting brain-teaser. All we can say for certain is that the catastrophic predictions of the past, with their scenarios of starvation and famine, have turned out to be wrong. Until now, at least…

Contemporary China has broadly succeeded in supplying the food needs of its huge population, even if there are still pockets of undernourishment in rural zones. The trouble is that these food needs are changing fast, as people become urbanized and aspire to greater diversity.

What will happen when they all start to put away as much meat as the Americans or the French? Lester Brown, president of the Earth Policy Institute, has calculated that in line with China's current economic growth, meat consumption per capita will match that of the US (125kg per annum) in 2031. Total meat consumption will have tripled by then to reach 180 million tonnes, that is, four-fifths of total world production as it stands today. Can Chinese agriculture possibly rise to such a challenge? What might be the impact upon the world grain market? We are in no position to answer these questions as yet. But producer countries are girding themselves for major shocks. The threat of a global food crisis seems to be already upon us, as outlined below. As we will see, our society of plenty, marred by the world obesity epidemic, is built on terribly frail foundations. In Chapter 11 on obesity and climate change, we cover the impact of meat consumption in more depth.

Will the world food price crisis curb the obesity pandemic?

The soaring prices of wheat, rice, maize, cooking oil and other food commodities began in 2007, and by 2008 had become a major preoccupation for governments and consumers alike, who had grown used to ever-falling food prices over the previous two decades.

Food riots have become widespread. From a health and nutritional point of view, the principle cause for concern is the increasing number of people going hungry and suffering from malnutrition in the poorest countries. Indeed a chorus of voices have proclaimed the need for immediate measures to be implemented, such as providing a safety net for the most vulnerable. But the question is also being asked about whether the crisis will have an impact on the prevalence of obesity in populations. We have very few documented examples that help us answer this question. The case of Cuba is the most well known, with a marked fall in the prevalence of obesity during the economic and food crisis of 1980–1990, followed by a return back to previous rising obesity levels (Franco et al, 2007). A different picture was seen in Congo during a period of economic crisis and structural adjustment programme (1986–1991): the prevalence of overweight (including obesity) in urban women continued to rise, but the number of people who were underweight also increased (Cornu et al,

1995). Therefore it seems that periods of economic crisis do not necessarily rule out the spread of chronic diseases, such as obesity and diabetes. Nevertheless, one would expect the increasing prevalence of obesity to be tempered by the economic situation in countries where expenditure on food represents an important proportion of income.

Concerning the current food crisis, uncertainty is rife about how long it will last. Will there be a return to normal? Will it be quick or not? Today everyone seems in agreement that prices will remain at a higher level than before the crisis; some even forecast the end of the era of cheap food, which would be a significant historic event. The effects could therefore be felt in the long term. But it is also possible that the consequences will be different according to the country and the population groups concerned: in rich countries, obesity touches poor communities in particular; and faced with less spending power, they could turn even more towards buying low-cost processed foods that are energy dense due to their high fat or sugar content. In low- or middle-income countries, it is families with a higher income that are most affected by obesity; and we could imagine that if high food prices remain in the future, it will slow down the nutrition transition and the spread of obesity to families with lower incomes in these countries.

Generally speaking, the food crisis, as well as the increase in oil and transport prices, could incite a number of countries to reinforce their degree of self-sufficiency: when it comes to food, local, healthier food systems could thus develop if these new agricultural policies become integrated with health policy that has explicit nutritional objectives. Whatever happens, the uncertainty about the effects of the current food crisis illustrates once again the need to implement or improve nutrition surveillance systems in all countries, to be able to monitor the health status of populations and take appropriate measures, whether they are for under-nutrition or overweight and obesity.

The issue of eco-costs

Production has not yet hit its theoretical ceiling, and so it can still be pushed up. There is even room for growth in the US, where some agronomists insist that crop yields can be made to double or even triple. But we need to know at what cost, both financial and environmental. For such mammoth increases can only be achieved, in already developed countries, through an even greater commitment to fertilizer, pesticides, irrigation, and the high-yielding crop varieties designed or engineered by biotechnologists (including GMOs).

Until now, steadily rising productivity has acted to bring down the price of food, enabling many impoverished countries to feed themselves for less. But it seems likely that this era has come to an end. If indeed there is continued pressure to increase production, we may see costs rise because of the products needed to fuel that extra output. Will consumers be able to keep up? And, given the insatiable thirst of intensive farming, will water supplies be able to do so?

The issue of 'eco-costs', or the costs to the environment of our production methods, is coming increasingly to the forefront. As far back as the 1980s, Professor David Pimentel of Cornell University worked out that 1kg of canned sweet-corn, worth 500kcal on the plate, used up more than 6500kcal to get there. It took energy not only to grow the plant, but also to harvest, process and package it, and more energy to transport and distribute it, and then to stock it, buy it and serve it. So pitiful a yield, for so great an input comes as a shock. Since then, other scholars have reckoned that it takes an average of 16 calories (4 'biological' and 12 'technical' calories) to produce 1 calorie for consumption. In that case, how come corn niblets are so cheap? Because the retail price does not pass on the all-important costs to the environment. Nobody pays for the damage done to ecosystems. It seems clear, however, that 9 billion people can never be fed if that kind of wastefulness continues. The choice is before us: either the industrialized world carries on pushing its high-performance agriculture regardless of the unsustainable environmental costs, which the planet can only bear on condition of keeping all other countries in a state of underdevelopment; or we look for another way. In any case, present yields will not be maintained for long if ecosystems continue to deteriorate. Our obese populations seem to be a product of a planet in dire straits.

Box 4.3 English apples, a storehouse of energy!

An apple a day keeps the doctor away. But depending on where it comes from, it may not be so healthy for the environment. This was demonstrated by Andy Jones, a researcher in the biology department of the University of

York, when he decided to look into the energy consumption behind apples available in the UK (Jones, 2002).

Currently, more than four-fifths of apples purchased in Britain are imported. To reach our stores from France, Sweden or New Zealand the crates of fruit must travel by sea, sometimes by air, before being trucked to retail outlets by road. What additional environmental costs might this supply chain represent, compared to an essentially local system of the sort that held sway in this country a few decades ago? Andy Jones set out to calculate the 'food miles' in relation to two market destinations for apples: Denbigh, a small market town in the north of Wales, and Brixton, a mixed neighbourhood in Greater London. His findings were startling. To transport a kilo (about 2lb) of New Zealand apples over 23,000km to shoppers in Brixton consumed over 10 megajoules of energy – about 2390 Calories (kcal). It took nearer 18 megajoules to get the same apples to Denbigh, equivalent to the energy used by a 100-watt bulb left burning for more than 2 days. Thus it turns out that the energy consumed in transport is 35 times greater than that expended, in the form of fertilizer, to grow the apples, and 9 times greater than the energy (calories) contained in the apples themselves.

Of course, this is an extreme, but genuine, example. If we compute the costs based on the true origin of the various tons of imported apples, the average transport-related energy consumption for one apple is between 4.6 and 6.5MJ/kg for Denbigh and between 3.5 and 4.5MJ/kg for Brixton. But this is scarcely peanuts where the environment is concerned. Andy Jones also estimated that a more localized production would cut greenhouse gas emissions by 87 per cent in the case of Brixton, and by 96 per cent in that of Denbigh. The author concluded that fresh produce sourced within a radius of 40km around the point of sale would considerably cut down the environmental costs. What is more, by planting a combination of different varieties, apples could be picked almost all year round. This is the option that UK agricultural policy deliberately rejected. Between 1980 and 1990, imports by air of fruit and vegetables almost doubled, and 240 per cent more fish was flown in. Air freight consumes ten times more energy than road haulage, which in turn consumes six times more than shipping by sea. At the same time, the distribution giants, which were responsible in 2002 for marketing approximately 77 per cent of all fresh produce sold in the UK, set up networks of just-in-time distribution centres based upon an energy-guzzling road transport system. As a result, the energy expended in the UK on food transportation represents close to 8 per cent of our total annual energy consumption per head. In 1992, imports of animal feed and related products alone, by air, sea or land, swallowed up 1.6 billion litres of fuel. The final reckoning for the planet will be no picnic.

Notes

1   Association nationale des industries alimentaires (ANIA).
2   The Green Revolution failed to bear fruit in sub-Saharan Africa, doubtless because yields can only increase if certain conditions are present, such as irrigation infrastructures, particular types of soil, and an auspicious political framework, all of which were lacking across the region.

Chapter 5

Welcome to Wal-Mart

The French press spoke with one voice when, in July 2005, there was a rumour that the American corporation PepsiCo was plotting to buy up the food giant, Danone. The entire political class lined up to defend one of the jewels in the nation's industrial crown. Shoulder to shoulder with the government for once, the Socialist opposition protested loudly. It warned the shareholders of PepsiCo that 'the most extreme reactions might ensue' were the American multinational to assume control of Danone, for 'many French people would take very badly what they would regard as a direct attack upon their identity'. Unsurprisingly its turnover of more than €13 billion in 2004 and its 90,000 jobs ensures that this company looms large in both the French economy and the national self-image.[1]

Industrial Meccano

Originally specializing in dairy products, Danone-BSN followed in the footsteps of its Swiss counterpart Nestlé by branching out over several decades to absorb a host of other brands. It is now a giant corporation that can offer consumers an ever-broadening choice of mineral waters, biscuits, ready meals and fresh produce. The Anglo-Dutch group Unilever followed a very similar path. Although in the early 1970s it was chiefly in the business of soaps and edible fats, today it controls dozens of prepared foods, desserts, baked goods and grocery brands, as well as cosmetics and personal care products. This strategy of integration, but especially diversification, was adopted in Europe as well as in the US, to build powerful multinationals with the capacity to dominate the market in grain, meat, dairy products and every other kind of foodstuff, while also operating the manufacturing and distribution processes, so that the same companies are involved at every stage, from field to supermarket.

Closer to the farming end of the chain than to the consumer are corporations such as Cargill, an American concern that having already dominated the cereal trade, is currently turning itself into the world's leading supplier of food commodities. After acquiring some Brazilian sugar companies, during the year 2005 alone, it snapped up chocolate factories in Europe, a vegetable oil refinery in Russia, olive oil presses in Italy and the largest sunflower oil plant in Romania, where months earlier Cargill had also bought the Corn cereal company, whose grain silos represented 10 per cent of the country's entire storage capacity.

As part of this giant industrial Meccano set that is reshaping the agriculture and food sector around huge multinationals, peasants in the more developed countries have completely redefined their own ways of working. First they became farmers, then agricultural production unit managers, gradually losing their independence as they became locked into actual chains of industrial production. Most of them no longer take their vegetables and free-range poultry down to the village market. Their job is to supply inexpensive but perfectly standardized and graded merchandise to international companies which have become, so to speak, their employers. Farms have turned into businesses producing raw materials, links in a long chain driven by a logic that is primarily industrial and financial.

This phenomenon has been taken to extremes in the US. Since 1960, the total number of farms has fallen from 3.2 to 1.9 million. But their average size has grown to 178ha, and their productivity has shot up by about 80 per cent. European farmers are small fry in comparison, with their trifling 18ha holdings.[2] This is what prompted the Irish Prime Minister, Bertie Ahern, in September 2005 to argue for the maintenance of the Common Agricultural Policy. Without these subsidies, he said that 'European farms on the margins of commercial viability would go out of business and European agricultural production would fall rapidly.'

In the name of higher profits, modern US farmers only produce, on the whole, a limited range of crops and livestock, principally maize, poultry, eggs, soya, beef... And they have entered into partnership with powerful corporations, in which the same company oversees every step of the production and distribution process. The corporation supplies the fertilizer, the seeds – no rival brands allowed – collects the harvested products, processes them into ready meals, and may often also distribute these to end users. What are the advantages for the company? Lower costs and the right to pocket the profits accrued by each link in the chain. Selling raw maize is already a good money-maker, as they go, but the market is not infinitely expandable. Whereas microwave-ready popcorn, more expensive to buy, is a much more lucrative proposition – provided one has access to vast supplies of maize. This explains why the corporations are so keen to sell the kinds of processed foods, including ready-cooked meals that afford the highest possible added value.

Box 5.1 Is obesity a question of price?

Why did obesity suddenly soar, in the US, during the 1980s? After all, between 1960 and 1980 obesity prevalence only rose by 2 per cent, to touch 15 per cent of the population (the current level in France). So why did it then double over only 25 years? According to an American research team (Finkelstein et al, 2005), the reasons are basically economic. The relative price of energy-rich foods (containing added sugars and fats) fell significantly during this period, while that of vegetables, fruit, fish and milk went up. These price differentials are due to the various technological developments that now enable products stuffed with energy to be cheaply mass-produced. The conclusion is that obesity is not merely a health issue; it is just as much the outcome of an economic system, in which the technology and organization of production make it more lucrative (and thus more 'rational', from a strictly economic point of view) to market and consume 'obesogenic' foodstuffs.

This model of industrial integration, having redrawn the landscape of US agriculture, is beginning to spread around the world. It is taking root most notably in developing countries, largely through direct investment by multinational companies that are mostly European and US based. Since the 1980s, more and more capital has been invested in developing countries for the production and distribution of food. Along with investment, comes heavy pressure to design products with high added value; to lower costs, improve efficiency and conquer new markets (Hawkes, 2005). The amount of US capital directly invested abroad in food processing operations rose from $9 billion in 1980 to $36 billion in 2000. Sales in those same countries followed suit, rising from $39 billion in 1982 to $150 billion in 2000. Since the late 1990s, US companies have been investing nearly $55 billion a year in food production and distribution systems abroad. The reasoning is clear: it's about gaining a foothold in rapidly growing markets, at the same time as taking advantage of the low costs of local production.

The bulk of these investments goes into highly processed products, for example sweets and soft drinks in Poland (more direct investment was poured into these two sectors during the 1990s than into meat, fish, flour, pasta, bread, sugar, potatoes, fruit, vegetables, vegetable oils and animal fats combined) and instant noodles, sweetened drinks, snacks, biscuits and fast foods in China. In return, the investors overhaul local food supply chains to increase the share of processed goods, for example ready meals and packaged foods in local diets. Sales of processed foods in low- or middle-income countries are booming, to the tune of an extra 30 per cent per annum. Precooked dishes, fizzy drinks, hamburgers, biscuits and ready-made desserts are flying off the shelves in Brazil and elsewhere in Latin America, as they are in Eastern Europe or in Asia. The tidy sums that have been invested in fast-food chains have multiplied the number of fries that are guzzled all over the world, and fuelled ruthless marketing campaigns, mostly targeted at the young, in a bid to dominate each market as early as possible.

Peasants out of the loop

Holding out against these potent trends, peasants around the world continue to work the land in the traditional manner, hoeing away at their tiny patch, operating within small family-based concerns. But on a global scale they submit to the system, rather than influencing it. In order to survive, many such 'workers of the land' in developing countries end up as field labourers in the employ of one or another multinational corporation that has been gradually buying up all the arable land in the region.

That is what happened in Kenya, for example. Between 1969 and 1999, the production of green vegetables more than doubled as a result, and exports were up by 6 per cent. Most of these exports are destined for the British market, under the control of the major supermarkets, whose average mark-up hovers around 45 per cent; but local farmers receive only 17 per cent (Millstone and Lang, 2003). By selling the vegetables that were grown so cheaply in Africa at a high price in England, the potential for profit is great for the distributor. But there's a hitch: nutritionists have observed that during the same period, the consumption of green vegetables in Kenya fell by almost 30 per cent. As the export economy raises the price of goods at home, small growers find themselves in the paradoxical position of being unable to afford the harvests of their own fields. Therefore if developing countries – like developed ones – are not consuming enough fruit and vegetables, it is not necessarily because they don't produce them.

Large-scale distribution takes off

The example of Kenya highlights the heavy influence that large-scale distribution is now beginning to exert upon the whole of the world food system. By 2002, according to the FAO, 30 leading distributors, for example Wal-Mart, Carrefour and Tesco, already controlled one-third of the world's food supply, with a joint turnover of $930 billion. This gives them phenomenal power, concentrated in ever fewer hands. The day is nigh when just five or six of these giant multinationals might be running the world food trade between them. In Europe, the market share of the ten leaders in the field is expected to rise from the 37 per cent they owned in 2000, to 60 per cent in 2010 (Lang and Heasman, 2004). This galloping trend has the ricochet effect of spurring more of the same across the entire food industry. Big distributors such as Carrefour or Auchan in France have an unbeatable ace up their sleeve: direct contact with the consumer. And it's no use producing if you can't sell the product. Whoever controls the last link in the chain has the power to make or break a product, one way or another. This is what makes the battles over large-scale distribution so ferocious, pitting monster companies against one another in the global ring.

In June 2005, the French group Carrefour owned 6680 stores all over the world (805 superstores, 1499 supermarkets and 4013 discount stores) and employed more than half a million people. Even so, it only ranks second in the distributor stakes, and by quite a long way. The all-round winner is the colossus known as Wal-Mart. A multinational that not only runs the world's largest department store business – it is also the world's largest public corporation, period. It has even overtaken General Motors, the automobile giant that symbolized American industrial muscle for much of the past century.

Wal-Mart's history is a classic rags-to-riches success story, of the kind the US loves. In 1962, Sam Walton (who died in 2005) opened his first dime store in Bentonville, Arkansas. The business thrived, allowing Walton to expand across the state and beyond, opening an outlet in Kansas, more in Louisiana, Missouri and Oklahoma in 1971, going on to Tennessee in 1973, until little by little Wal-Mart had spread all over the US. In 1991, the company went international, moving first to Mexico and eventually expanding into Puerto Rico, Argentina, Brazil, Canada, China, South Korea, the UK and Germany. Today Wal-Mart employs some 1.3 million people to run more than 5000 stores worldwide (most of them in the US), with a turnover of $265 billion – roughly equivalent to three-quarters of the French state budget! More than 100 million Americans, almost a third of the total population, shop at Wal-Mart every week.

Sam Walton's secret was simple: he offered discount. He sold at lower mark-ups in order to achieve higher sales volumes of a wide range of lines, from clothes to toys to food. And he made his bet on a certain rural America where purchasing power was modest, but people had done well all the same from the economic upturn of the 1960s. Wal-Mart's long-lived slogan summed up its basic philosophy: 'Always low prices' (recently changed to 'Save Money. Live Better'). Wal-Mart slashes prices and lets everyone know it. And the consumer is delighted. But this strategy has an enormous impact on the system as a whole. Firstly on suppliers, who are relentlessly leaned on to offer Wal-Mart a better deal. And the company has no qualms about outsourcing its supplies (from China especially) to take advantage of the best deals on the market. Secondly on Wal-Mart's employees, who are notoriously badly paid. Lastly, its strategy hammers the competition, forcing other big distributors to match Wal-Mart's prices if they wish to stay in business.

The supermarket Eldorado of emerging nations

Having largely run out of scope for opening any new stores in their native countries, the large supermarket groups have for some years now been transferring their attentions to foreign markets. They have launched new operations in populous countries like China and Brazil, where the appetite to consume is on the rise.

Carrefour opened its first Chinese store in 1995. Since then, the French corporation has set up 60 more in that country, and plans were afoot for another 10 or 12 in the capital, Beijing, by 2008. In Brazil, the Carrefour brand is already top dog, controlling 12.6 per cent of the market. And yet in 2005 the company decided to sink €200 million into 17 new superstores, on top of the 84 it already possessed in that country. Its subsequent plans were to expand at a rate of 15 stores every year for the foreseeable future.

China and Brazil are not the only Eldorados of hyper-consumerism, and this type of store can be found all over the world nowadays, even in areas of deprivation. In Latin America, supermarkets accounted for 50 or even 60 per cent of all food shopping by 2000 (in Brazil, the figure was 75 per cent in 2001) (Reardon and Berdegué, 2002), as compared with just 10 or 20 per cent a decade before. In the space of ten years, the dynamics of distribution have changed beyond all recognition, forcing the food system as a whole to adapt to the needs of the multinationals, at the expense of traditional grocer's shops and street markets. Nicaragua, the poorest country in Central America, was already host to 60 supermarkets by the end of 2002.

From being confined to a handful of select capital cities during the period between 1960 and 1980, these chain stores fanned out to occupy smaller cities and towns, even in remote regions (around one-third of small towns in Chile and up to 40 per cent of those in Costa Rica now boast a supermarket, sometimes two). From there they jumped into low-income neighbouring countries: from Costa Rica to Honduras and El Salvador during the early 1990s, and from Chile to Peru, Ecuador and Paraguay. And if their outlets were at first located in wealthier neighbourhoods, by the turn of the century they penetrated the working-class districts, where they implemented aggressive hard-discount strategies.

Just as in the US or Europe, these supermarkets were an answer to real popular demand. After all, Latin America has, like Western countries, become overwhelmingly urbanized (the proportion of town-dwellers in Chile rose from 75 to 86 per cent between 1970 and 2001). As more and more women enter the workforce, they too find themselves pressed for time and don't wish to spend it in the kitchen; the pattern is repeated in all emergent countries. Besides, any increase in disposable income tends to enhance the desirability of processed foods. Refrigerators have become ubiquitous (50 per cent of Chilean families owned one in 1987, rising to 82 per cent in 2000), so that perishables can now be bought in bulk and kept fresh for a couple of weeks, something that was impossible before (a Kenyan study has shown that the chief determinant of fruit and vegetable purchase in supermarkets is access to a fridge). Finally, as we saw in the case of industrialized countries, eating out is becoming a habit. In Argentina, for example, 18 per cent of total expenditure on food in 1996 was lavished on meals away from home, against a scant 8 per cent in 1970 (Ghezan et al, 2002).

On the supply side, the liberalization of world trade, which took off in South America at the start of the 1990s, facilitates imports and enables these to be acquired on a massive scale, a bias that works in favour of the largest distributors.

The bigger fish thus swallowed up the smaller, one by one, until very soon only the sharks were left. This process set in during the late 1980s, when national chains began to go after independent retailers. The Argentino-Uruguayan chain Disco, for example, having absorbed several Argentine distributors, bought the Chilean grocery chain Santa Isabel; it then went into partnership with the Dutch powerhouse Royal Ahold, until this in turn took over Disco and all of its subsidiaries.

As the 1990s progressed, a second, even more predatory wave broke with the arrival of US and European enterprises. Western companies bought up numerous successful local chains, before turning their attention to smaller distributors and independent stores in the course of tightening their grip on the market in the poorer hinterlands. La Fragua, the leading supermarket in Guatemala, joined forces with Ahold in 1999. Two years later the chief Costa Rican distributor, CSU, merged with La Fragua and Ahold to form the Central American Retail Holding Company (CARHCO), which thus came to operate 363 stores worth US$2 billion in sales across Central America by 2004.

The outcome of all this: in Argentina, five major supermarkets controlled over three-quarters of the groceries sector in 2001. The same companies had 80 per cent of the market in Mexico, 85 per cent in El Salvador, 96 per cent in Costa Rica – and a thumping 99 per cent in Guatemala. Only crumbs were left over for the small guys.

Africa, too, is becoming scattered with supermarkets. In Kenya during the 1990s, their importance in food trade networks was negligible; by

2003, however, they commanded as much as 20 per cent of the retail business in the capital. Shortly after this they began branching out from Nairobi into small and medium-sized towns. This doesn't mean the country was getting richer: contrary to a widespread assumption, over half of the customers of Nairobi's supermarkets are low-income consumers, and yet their spending accounts for 36 per cent of the turnover.

Food choice: Is it real or sham?

The proliferation of supermarkets and superstores contributed to the downward pressure on food prices. This was, after all, the original purpose: to reduce costs by cutting out the middlemen that stood between producers and consumers. So has this been all to the advantage of the consumer? Not quite. True, supermarkets have done a lot to encourage, for example, higher intakes of dairy products all over Latin America by providing long-life milk, yogurts and desserts. And a study carried out in Detroit, in northern US, found that Afro-American women who shop in supermarkets eat more fruit and vegetables than those who shop elsewhere (Zenk et al, 2005). It is also true that the range of products in large stores is astounding at first sight, appearing to offer more choice than ever before. But behind this extraordinary diversity, there is little information for the consumer about what really goes into food. Among the many kinds of new, fun cereals aimed at children, for example, which might be the best for maintaining a balanced diet? It's very hard to know, even by scrutinizing the labels. And the existence of many different brands doesn't necessarily imply any real difference on the nutritional level: endless rows of products that are all equally rich in fat, sugar and salt suggest that beneath the appearance of choice lies an essential sameness. To swap crisps for chips, or to alternate between cakes and puddings, does not mean that one is varying one's diet. It's possible to buy a lot of seemingly different products and yet always be eating the same thing, nutritionally speaking. To date, there are remarkably few studies that might help to assess the actual impact of supermarkets and superstores on the nutritional value of our meals. It is safe to assume, even so, that the relentless insistence of large-scale distributors on sourcing supplies at rock-bottom prices do not give producers much room to be fussy about the nutritional virtues of their products. There's no advantage for the farmer in raising leaner beef, for example; on the contrary, he risks losing competitive edge.

The fact that supermarkets are increasingly aimed at low-end consumers should also be kept in mind. For while one could easily put together a healthy, varied menu from the huge choice on offer in such places, people on tight budgets are usually unable to make the most of it. They get into the habit of buying pretty much the same things over and over again, gradually losing their sense of its nutritional value. This prompts us to wonder whether supermarkets contribute directly to obesity. Are they not, on the contrary, simply keeping step with it? It's a question worth asking.

The levelling of culinary cultures

The rash of supermarkets and fast-food chains breaking out all over the world certainly seems to have contributed to one conspicuous development: the homogenization of food cultures. Whether you're in Paris, Cairo or Brasilia, you'll always find an identical Big Mac with its unvarying cola drink in the shopping centre attached to the superstore where you could have filled your trolley with almost exactly the same products, give or take a few.

Local culinary traditions are fast dying out as a result, like so many endangered species. For centuries until now, each culture had evolved its own ways of eating, derived from a balance between the local environment and historical or cultural factors, gained from experience. In North Africa they came up with couscous, a dish based on pulses and cereals, garnished with mutton. In Crete they developed the famous 'Mediterranean diet', rich in fruit and vegetables, cereals and olive oil, plus a little wine. The available ingredients were more restricted in Nepal, giving rise to a less sophisticated diet whose energy comes 80 per cent from grains (compared to 20 per cent in the US). Japanese cooking relies greatly on raw fish, whereas the Chinese based theirs on the staples of rice and soya.

As we have seen, the food transition unfolding in almost every developing country turns people away from their ancestral cuisines and towards foods packed with added sugars and fats. Superstores and fast-food outlets, with their ability to deliver the same standardized and affordable product lines in all the great cities of the world, are clearly not the only factors to blame for this situation. But they have done their bit; that much is certain.

Television, particularly through its commercials, has played an equally influential role. When television first appeared in Senegal, during the 1970s, it broadcast for only a few hours a day, and the programming consisted of little more than news and local reports, sprinkled with the odd commercial break. Then the Nescafé brand stepped in to sponsor a weekly movie, preceded and followed by several minutes of Nescafé adverts. Thus all the TV sets in the country, most of them watched by dozens of villagers at a time, were turned into powerful vectors for the brand. Ten years later, Senegalese TV was on air for much of the day and evening, and the advertising was relentless, with particular stress on everyday convenience foods like instant coffee and stock cubes. These commercials devised and produced in Abidjan, capital of the Ivory Coast, promoted a glossy image of consumerist family life based on the model of a wealthy Abidjan family, living in a spacious detached house – the situation to which most Africans doubtless aspired. Today, the same channels broadcast a loop of programmes and commercials of the kind that can be seen all over the world, all preaching the same lesson of Western consumerism.

One may be tempted, in light of the above, to perceive nothing but the negative aspects of globalization. On one side, the bland goods foisted on the world by mighty multinationals, on the other, the exquisite products of the local *terroir*. But things are not really so black and white. First of all, globalization has enabled many people to enjoy what were previously unknown flavours and dishes, as the delicacies that were once confined to their native regions found fame beyond national borders and overseas. We can have sushi and tacos in London, just as the Mexicans can sample choucroute. This is not a recent phenomenon. Different kinds of foods have always travelled around the world, on the shoulders of exploration and conquest. And the much-vaunted 'local products' were often themselves introduced from somewhere else, or were composites making novel use of alien ingredients. No one would claim that pepper, say, is a native European plant. As for other 'traditional' dishes, it turns out that many were invented less than 100 years ago. Finally, for the sake of irony, consider what has lately become one of the most traditional snacks of all in the US, courtesy of the fast-food habit: none other than 'French fries', the classic French *frites*, that migrated to America in the luggage of GIs going home after the Second World War!

History is there to remind us that culinary cultures are never set in stone, for societies are constantly reinventing the ways in which they eat, and it may be that the pressures for change brought to bear by the contemporary food industry are no heavier than those applied in centuries past. Creole food, for example, was born from the intermarriage of African, European, Indian and Far Eastern cooking. Our familiar breakfast drinks of tea, coffee and cocoa have also come from far away.

Maize, the staple food of the ancient civilizations of Central and South America, was introduced into Europe after the discovery of the New World. It was originally dubbed 'Spanish wheat' or 'Turkey wheat', and was stigmatized for a long time as pauper's food, fit for those who couldn't afford to buy real wheat bread. The Italians made it into *polenta*, the Romanians into *mamaliga*. The new crop was brought into the Congo and Angola by Portuguese traders in the 16th century – Kikongo speakers called it *masa mamputo*, 'grain of the white man' (McCann, 2005), and local tribes rapidly incorporated it into their diet. Variations on its use in Africa today include *mawe* or *ogi*, a kind of fermented cornflour; *akassa*, a thick mash of fermented flour baked in leaves; *ablo*, a steamed ball of leavened maize; and *aklui*, a kind of couscous made of corn and boiled into porridge.

But the fact remains that the evolution taking place today, in the framework of globalization and mass urbanization, belongs to a different order. The earliest global exchanges had introduced Western countries to new plants and crops originating in the South and East. But the wind is now blowing the other way, and the dominant tendency is to flood the whole planet, that is, to transfer to an unprepared South the processed foods and corporate brands of the North. In this sense, the proliferation of fast food and soft drinks is the most telling sign of the new food system. For the modern way of eating undermines traditions of home cooking, with its industrially produced convenience food, its one-size-fits-all formula. Thus local populations have fewer opportunities to use new ingredients, a basis for inventing their own recipes. Pizza, which has emerged as the most popular dish in the world, is virtually indistinguishable in all of its forms, whether served up at a Pizza Hut in New York or delivered to your door by a firm in Hong Kong!

Old and valuable skills are thus potentially threatened. Such as the techniques of Amazon peoples for detoxifying manioc to make it safe to eat, or the *chuño* process developed in the Andes mountains for the efficient preservation of potatoes: the peasants expose the tubers to the nocturnal frosts, then crush them underfoot in order to expel most of the water, before drying the flesh and storing it for future use. In Asia, various soybean fermentation techniques were refined over the centuries giving rise to quite different end products, from the *tempeh* of Indonesia and Malaysia to Japan's *miso*. Another soy product popular in Japan, perfected by more than a thousand years of refinement, is *natto* – obtained by cooking the beans and fermenting them with a pinch of the bacteria *Bacillus natto*, which lends the stuff its sticky, stringy texture. In West Africa, women ferment nere seeds from *Parkia biglobosa*, a leguminous tree also known as the African locust bean, to obtain a savoury condiment that goes by the name of *netetou* in Senegal, *iri* or *dadawa* in Nigeria and *sumbala* in Mali. Tens of thousands of tons of it are consumed every year, for its flavour reminiscent of meat.

But these fermentations were not only developed to vary the pleasures of taste. They also fulfil important nutritional functions, by making leguminous plants more digestible and providing useful amino acids (*natto* and *netetou* are composed of 40 per cent protein), vitamins (thiamine and riboflavin) and essential fatty acids. How many such specialities will survive the inexorable advance of fast foods and supermarkets?

Is there a direct correlation between this loss of culinary diversity and the rise of obesity? We suspect there is. Because good health is to some degree a matter of cultivating a balance with the natural environment, and many of these vanishing culinary traditions had been closely adapted to local conditions. The fact that such diverse human groups exist upon the Earth, whose natural diets are so unlike one another's, is surely the sign of their adaptation to very different climates and conditions. Until very recently, a Mayan farmer would never have eaten the same foods as an Inuit hunter. To ignore these specificities is to invite new patterns of food behaviour that cannot be other than ill-adapted. Especially if they involve embracing a diet high in fat and sugars, just when energy requirements are in free fall among almost all the peoples of the world.

Notes

1   Whether or not as the result of this posturing, the buyout did not happen after all. It had probably never been seriously envisaged by PepsiCo's executive board. Nonetheless the story illustrates better than most the growing importance of these multinational food companies, in both the French and the global economies.

2   Average for the 15 European member states, before the EU's enlargement to 27.

Chapter 6

Culprits or Scapegoats?

Super Size Me, the controversial film by Morgan Spurlock, made a simple claim: that America has grown obese because of McDonald's. It was also somewhat simplistic. This chain of so-called restaurants provides an obvious scapegoat for anyone looking to pin the blame. All the same, nutritionists have long emphasized the important contribution of the food industry to the flood of calories that has engulfed consumers over the last few decades. To what extent is the industry truly responsible for the world's obesity epidemic? We have seen its economic value to countries like the US or France. We have traced its development as it became highly concentrated, then globalized, then reorganized around large-scale distribution. Is it scientifically possible to disentangle the various responsibilities of these huge industrial and financial companies? Can the obesity epidemic be regarded as the inexorable consequence of our production and distribution system?

Neither saints nor sinners. The goal is profit

It is time to recall something that should be self-evident: the food industry never deliberately set out to make people fat. Just as one might reasonably assume that the tobacco industry doesn't go out of its way to induce the cancers that smoking unfortunately causes in certain people. In both cases, these outcomes are calamities that the boards of the corporations concerned could happily have done without – even if only for the sake of the billions of dollars that cigarette manufacturers had to fork out in compensation.

It is clear, instead, that the food industry – like the tobacco industry – is built around companies whose prime objective is to make money. Like it or not, such is the rule of the game in our capitalist societies. Agri-food companies are thus bound to produce, and sell, anything that's likely to reap them a legal profit. The long-term effects on consumer health only start to become of real concern to them if these effects threaten to jeopardize profits. To put it starkly, whenever a food company comes up with a new product, swimming no doubt in fat and sugar but easy to make and easier to sell, its first reaction is not to wonder whether this product really answers a nutritional need. After all, it is only giving consumers what they want. And what they want is novelty, affordability, pleasant taste, attractive appearance and a safe product. If people aren't that bothered about nutritional value, then it's not up to the company to bother on their behalf. In a fiercely competitive market economy, the company's job is to get consumers to buy more of its products, or at least to buy its brand in preference to another's. Should it ever neglect this duty, the shareholders would soon put it straight.

More nosh for the same dosh

So, how is the consumer to be coaxed into buying more? In developed countries, the food industry was faced with a bit of a problem early on: people who have eaten their fill stop buying, because they're full. Oh dear! Something had to be done to correct such economically regrettable behaviour. The Eureka moment was the realization that feeling full was, in fact, a thoroughly subjective notion.

Why were McDonald's customers of the 1960s so satisfied with one little portion of fries? Why did they so seldom come up to order another? This conundrum began to bug the marketing strategists. Surely people could be made to eat beyond the mere satiation of hunger, if only one could press the right buttons. Yes, but which? One button was found during the 1970s (Ledikwe et al, 2005) and the resultant strategy was fully rolled out during the 1980s: it was all a matter of offering larger portions. Dumb, perhaps, but somebody had to think of it.

And to entice customers to buy these extra-large helpings, they were made to feel they were getting a bargain: lots more nosh for virtually the same dosh. The maker was not really losing out, since, as noted earlier, the cost of the actual food as a proportion of the price of a food product tends to be low – as little sometimes as 5 per cent! Thus to double the quantity of chips in a helping does not mean doubling the costs incurred. And even if the company does not make twice the profits on a double portion, it still makes more than on a single portion (since the price hike is always a little higher than the net cost of the extra food). As for the consumer, he's plainly getting more for his money, which is also a face-saver: opting for the 'maxi-size' choice makes him look like a smart and savvy consumer, instead of a pig stuffing himself on two portions of fries. A marketing masterstroke!

Servings began to get larger. Tentatively at first, then recklessly. So much so that today, the meals sold in American fast-food joints are between two and five times bulkier than they were 20 years ago (Ledikwe et al, 2005). The same goes for the units sold in shops, which have been conditioned into following the trend. Whereas during the 1950s, soft drinks and other sweetened beverages were sold in the US in 6.5oz bottles, by the 1970s the standard size had risen to 12oz. In 2000 it reached 20oz, that is, over half a litre. Crisps too are sold in packets that are up to three times bigger that they were a few decades ago. At the same time there has been an explosion of 'All-you-can-eat for X dollars' formulas in restaurants and other buffets in the same mould, in which it takes some willpower to resist the pleasure of refilling one's plate, seeing as it's all for the same price.

Now, is this food, bought and sitting on the plate, actually eaten up? Sadly, it is. Studies show that after infancy, our appetites adapt to the volume on the plate. For example, when adults are offered four different-sized portions of macaroni cheese, their calorie intake is 30 per cent greater when they tuck into 1kg of the stuff than when they consume half that amount, hefty though it is already (Ledikwe et al, 2005). In other words, a full stomach can be filled further still, almost unknowingly. The same phenomenon was observed whether the experimenter served the subjects or whether they were free to select their own portions.

Worse still, the fact of having eaten more did not make anyone feel more replete. Herein lies the paradox: the subjects who ate the largest portions felt no more 'full' than those who ate the smallest. It transpired at the end of the study that more than half the participants had not even noticed that the portions served were of different sizes. This is not as improbable as it sounds. A good number of studies show that consumers find it hard to assess the size of food helpings that do not conform to a predefined shape, such as pasta or rice dishes. And the larger the helpings, the harder they find it, so that people become incapable of estimating quite how much they have ingested. They are therefore prepared to ingest however much the provider decides is good for them. Similar studies have been carried out using different-sized sandwiches, packets of crisps, tubs of popcorn and so on. All of them confirm the same hunch: the more food someone is offered, the more they will eat.

The quantity of food thus downed is independent of its calorie count. That is to say, you will polish off a large bowl of chips exactly as you would the same large bowl of salad. Tests show that we eat as much as is put before us. Americans, who have increasingly grown into the habit of eating away from home, where ever larger portions of sugary, fatty, processed food are put before them, have been licking the platter clean with no idea that they were eating so much more at a sitting. The conditions are therefore in place for maximum calorie intake. The French, by contrast, by holding to the principle of family meals at home, have been relatively shielded from this phenomenon.

Aware of having pushed things a little too far and anxious to avoid costly lawsuits, the food industry is beginning, in the US at least, to temper the logic of 'more is more'. PepsiCo, for example, recently announced a decision to reduce the size and total calorie count of the items it sells in American schools.

The power of advertising

The fact that a considerable number of products are 'obesogenic' is old news. Nutritionists have been sounding the alarm for decades. But in keeping with the economic model that began to prevail from the early 1980s, the manufacturers – some of them in good faith, no doubt – relied on the market to take care of the problem. After all, consumers would surely opt for the healthiest products, and a product that nobody buys is commercially dead. The responsibility, then, lay firmly in the shopper's court: it was up to them to make sensible choices! And as they did so, the most unwholesome offerings would automatically be condemned to fail. This optimistic vision was at best naive, for it completely overlooked the persuasive power of marketing and advertising.

For anyone still harbouring any doubts about it, here is what the WHO (2000) had to say about advertising in Britain: '£86.2 million was spent on promoting chocolate confectionery in the United Kingdom in 1992 compared with only £4 million spent on advertising fresh fruit, vegetables and nuts.' In other words, 20 times more money was spent that year on promoting confectionery in the UK than on marketing fruit and vegetables. It cannot seriously be claimed that the contest between foods that are good for you and those that are not is held on a level playing field.

Kids, the privileged target of the Big Five

Television plays a decisive role in this game. It can particularly influence children, since they are exposed to more food commercials than adults are; the next most relentless advertising blitz directed at kids is for toys. One study (Dibb and Castell, 1995) found that seven out of ten commercial slots in peak viewing hours for children were promoting food. And every one of these food ads was pushing products rich in fat, sugar and salt. The WHO (2000) has drawn attention to the same phenomenon, reporting that '91% of foods advertised during peak children's viewing time in the USA, and a similar proportion in the United Kingdom, were high in fat, sugar and/or salt'.

The vast majority of such commercials are for products known in the trade as the Big Four: breakfast cereals with lashings of added sugar, soft drinks, confectionery and savoury snacks. A fearsome foursome of calories that was overtaken in recent years by a fifth villain, now the unrivalled leader of the pack: fast foods, whose advertising budgets have soared. McDonald's, which in 1990 was already ranked as the fifth-largest advertiser in the world, became the second-largest only two years later (Horgen et al, 2001), and reached the number one spot in Europe by 1997. The gang has now been dubbed the Big Five by nutritionists.

Ads for these five categories of foods displaced those promoting basic essentials such as bread, fruit and vegetables, first in the US, and then throughout the world. Britain presents a particularly extreme case, as this unhealthy advertising had dominated children's TV by the late 1990s. Can it be a coincidence that this was also the decade when obesity in British children began to get out of hand?

Fun and games...

In order to convey the best possible image of themselves, the Big Five lay more stress on entertainment and taste than on goodness and nourishment value. As can be seen from the most cursory glance at such messages, the big selling point is the concept of 'fun'. Cool, amusing characters are order of the day. And oddly enough the child actors used are always slim, fit and bursting with energy, while the stuff they are eating is, to put it kindly, 'of scant nutritional value' (Byrd-Bredbenner and Grasso, 2000).

Does such advertising really influence children's ideas about healthy eating? Different studies have come to different conclusions on this point. It has been shown for example, that adverts for soft drinks may impair the ability of younger children to determine whether certain products contain real fruit or not.

Of course, advertising chooses its words with care. Cereal ads often tell you that the product is beneficial 'as part of' a balanced breakfast, without going so far as to claim that it constitutes a balanced breakfast in itself. The devil is in the detail. But is the viewer alert to such nuances? Again, a product will usually be described as 'energy-rich' rather than 'sugar-rich', and it certainly sounds better to say 'They're Grrreat!' than 'They're Fffattening!'. Vague terminology on the lines of 'contains natural fibres' is advantageous in that it is non-committal (the consumer is bound to find a couple of these if looking really hard), while giving the impression that this is actually a health food. Thus such terms are liberally mixed in. Close to half of all adverts for food and drink contain wilfully ambiguous or downright false information of this sort (Byrd-Bredbenner and Grasso, 2000). Sometimes it is a question of misleading comparisons: such-and-such a chocolate bar is said to be as good as a glass of milk, or to contain as many vitamins as a piece of fruit. But while it brags of the similar vitamin content the script is unlikely to mention the difference between chocolate and fruit in terms of calories and other nutrients. No two ways about it: the only real equivalent to a piece of fruit is ... another piece of fruit.

Television commercials are not the only ones to play with semantics in such a way as to sow deliberate confusion. On the menu of a well-known chain of steak houses, a claim has appeared since 2004 to the effect that the bison meat it offers as an alternative to beef contains less fat than any number of fish, including salmon. Put like that, it may well be true. But the subliminal message – 'beef is even healthier than fish' – is open to question. A bison grazing freely on the plains has little in common with intensively-reared beef. Their diets are worlds apart, and so is the fat content of their flesh. Moreover, the text of the menu omits to mention that while salmon is certainly an oily fish, the proportions between the various fatty acids are completely different and so weighted as to present a correct balance between 'good' and 'bad' cholesterol. We cannot say the same of beef.

A similar confusion is cunningly maintained with regard to dairy products, which are undoubtedly 'rich in calcium' but also, more discreetly, in fats.

...will do the trick

Do such commercials really influence our children's eating behaviour? Unfortunately it appears they do (that's their purpose, after all). But it is hard to tell whether the blunt instrument of commercials exerts more or less of an influence than other factors. Among these must be counted the choices and behaviours of the parents, which have a decisive impact on what children actually eat: if Dad sits nightly in front of the telly with a mega-bag of crisps, or Mum regularly tucks into a tub of ice-cream, then the child is highly likely to do the same.

Nevertheless, studies show that advertising does influence primary-school children who are asked what their favourite foods are, or what they snack on in their free time. They also show that promotional messages on vending machines sway the students' choices, in one way or another: appropriate messages are just as likely to persuade them to buy healthier snacks. Either way, the admen are not wasting their time...

A 1990 Canadian study (Goldberg, 1990) compared cereal consumption among Anglophone and Francophone children in Montreal. In those days, Anglophone kids in Quebec mostly watched American TV while the Francophones tended instead to watch Québécois channels, where adverts targeted at children had been banned since 1980. The French-speaking youngsters were underexposed to cereal commercials as a result. The study found that the children who watched the most American TV were also those who ate the most cereal, regardless of household income band or home language. Other experiments have shown that when kids watch a lot of food adverts, they tend to pester their parents to buy specific products: more often than not, those with a high fat, sugar and salt content.

Commercials aimed at children exert two potent effects: the brand effect, inciting a child to demand one brand rather than another because they have seen it on TV, and the food category effect, which encourages children to desire more products of a particular type (chocolate bars, say). There is no doubt that advertising can encourage children to eat more junk whatever the brand, at the expense of other categories of food. The almost total absence of fruit and vegetables in children's commercial slots may thus partly explain their seemingly spontaneous lack of interest in their greens.

At the end of the day, is it fair to blame advertising for the rise of obesity? Certainly the link between watching television and having an unbalanced diet, one that invites cholesterol and weight problems, has been conclusively demonstrated. In 2005, for example, one study even showed that the proportion of overweight children in the US, Australia and several European countries was directly correlated to the density of television ads for fatty and sugary products (Lobstein and Dibb, 2005). Conversely, it has been shown that if limits are placed on an overweight child's TV time, they slim down considerably in a matter of months. Still, what is the exact nature of the link between TV and overweight? Are commercials fattening? Is it because kids are constantly nibbling while they watch? Or is it because lounging in a comfy chair is not the best way to burn off calories? No doubt the answer lies in the convergence of all these factors, in proportions that remain to be determined. Some studies suggest that the more commercials children see, the more they feel like snacking and the lower the quality of the food they eat. Others, however, would indicate that commercials are not as important, ultimately, as the attitude of the parents. Note that most studies focus only on the direct effect of TV ads on children, ignoring more tangential influences. Fast-food adverts, for example, may leave a child unmoved but inspire her parents to take the whole family to McDonald's. In this way the notion that lunch in a fast-food restaurant is a normal, not to say desirable, thing to do becomes reinforced within the child.

The authorities strike back

Whatever the real impact of these commercials may be, they have aroused official concern in several quarters. At the beginning of 2003, Britain's House of Commons Parliamentary Health Commission announced its intention to examine the role of advertising, as part of a raft of measures to tackle obesity. Its annual report a few months later recommended the adoption of the precautionary principle in this matter: despite the impossibility of proving beyond the shadow of a doubt that food adverts were responsible for the mounting obesity of the population, and in view of the high probability of some such responsibility, stronger regulation of commercial messages was called for.

A range of international organizations, including the WHO and FAO have arrived at the same conclusion. Since food advertising must incontestably be implicated on some level, there is nothing wrong with trying to moderate children's exposure to the marketing blitz. After ten years of heated debate, Sweden moved in 1991 to ban all TV advertisements aimed at the under-12s. Greece, for its part, opted for stricter controls over advertisement content.

In France, the food safety watchdog AFSSA was thinking along the same lines. The agency's view was that scientists are unlikely ever to be able to prove outright that advertising has contributed to the rise of obesity, because the issue is extremely complex and involves a multiplicity of factors. So it posed the question from another angle: in a world where child obesity is a daunting problem, how sensible can it be to let children be bombarded with ads to guzzle energy-rich foods? The precautionary principle advises us to err on the side of safety.

Picking up on this in 2003, a French member of parliament proposed a ban on all ads for foodstuffs that could by no stretch of the imagination be considered part of a balanced diet. The law would also compel manufacturers to include a nutrition education message, approved by public health experts, in any food commercials directed at children. But this bill was watered down under government pressure, so that advertisers were merely required to fund the creation and transmission of such nutrition messages, on the advice of the AFSSA, during the same time slots as their own commercials. This would be achieved by means of a levy on the marketing budget for their products. Even in toned-down form, the bill enraged the food industry that thought itself perfectly qualified to make its own ads with its own agencies, and refused to accept that it could not be both judge and jury in this matter.

The principle of a levy still stands at the time of writing, but was pared back to a very modest 1.5 per cent of advertising expenditure, and food advertisers now have the choice of two options: either pay the 1.5 per cent tax to the National Institute for Prevention and Health Education (INPES) to finance campaigns to prevent obesity, or accompany food adverts with information messages about health. The health messages are in the form of moving banners on TV and when they appear in newspapers/magazines they are similar to health warning labels on cigarettes, examples of the main messages are:

- 'For your health, eat at least five fruit and vegetables a day';
- 'For your health, engage in regular physical activity';
- 'For your health, don't eat too much fat, sugar or salt';
- 'For your health, try not to nibble between meals'.

It remains to be seen whether the sum collected will suffice to 'implement a genuine policy of public information with regard to the risks associated with poor eating habits', to cite the bill's original ambition.

Powerful lobbies

The howls of protest unleashed by this member's bill, tame as it was, made something very clear: as soon as it perceives a threat to its interests, the food industry will fight back tooth and nail. It has been very effective at creating muscular lobbies to defend its corner. The sugar industry in particular – whether in the UK, France, Cuba or Brazil – is tirelessly vigilant, and moves heaven and earth to prevent the publication of any reports linking sugar intake and obesity, or attempting to recommend maximum levels of sugar content in certain products. The jazzy ad campaign that ran on French TV during the autumn of 2005, which likened the authorities' concern to control sugar consumption to some totalitarian desire to repress the very concept of pleasure, was wonderfully revealing in this respect: some sections of the food industry will stop at nothing to keep the tills ringing.

To this end, there are many handy tools the industry can resort to. Donations to political parties and legislators, funding of academic conferences or journals, that are reluctant to bite the hand that so 'generously' feeds them. It's no secret that in the UK as in other countries, the research of many nutritionists is partly bankrolled by agriculture and food companies.

These lobbies proved their worth in May 2005, when members of the European Parliament resolutely supported the food industry against a group of consumers' associations seeking to regulate food labels. The European Commission had tabled a project to ban lollipops, for example, that are composed of more than 90 per cent sugar, from being touted as 'fat free'. Consumer groups who objected to the ambiguity of such descriptions (do they not imply that this is a 'lite' product, which can be enjoyed with no worries?) were demanding a clampdown. The text of the bill proposed that such labels could only be used if the product in question did not contain excessive quantities of fat, sugar or salt. But MEPs were having none of it and rejected the bill on 26 May. Some justified their action by saying that there was no such thing as 'good or bad products, only good or bad diets'. Others, however, hinted at deeper motives for their decision, arguing that the Commission's formula was 'anti-liberal' and would have imposed intolerable 'over-regulation' on companies, especially onerous for small and medium businesses. They could hardly have admitted more candidly that the health of a business takes precedence over that of its customers. For that's the problem in a nutshell: all regulations, no matter how important for public health, pose a threat to short-term profit. We are faced with a clear choice. Which shall it be, a more high-performing economy or citizens in better health?

Lobbies against lobbies

The activities of the various industrial lobbies have been exhaustively analysed by a number of authors already. Lately, however, a new and no less powerful kind of pressure seems to have entered the ring from the other side: that exerted by banking and insurance corporations. The world of finance is beginning to wake up to the hazards of a system that has spun out of control, and nervously anticipates a backlash – as in the case of the tobacco companies – that would oblige them to foot the bill for a new angry wave of prosecutions. That is why they are nudging the food giants to come to their senses, with increasing urgency. The major American investment banks are behind a spate of recent reports detailing the very real risks courted by the food industry if it does not do something about the composition of its products. The message seems to have got through to a number of brands, which are now coming up with new 'healthy eating' lines that claim to address the problems caused by an unbalanced diet. Pitting bucks against bucks, this counterblast could well do more than any amount of appeals to a sense of ethics, or the arguments of public health. The investment banks have wagered great chunks of capital on the food business, and they don't wish to see it go under. The insurance companies are in an even more awkward position, for they're the ones that would have to pick up the tab. Hence mounting pressure is being applied to the entire edifice of producers, distributors and, above all, fast-food chains – the most direct offenders – urging them to review their production and marketing practices in depth.

A genetic link?

But having said all that, perhaps this is simply a matter for our genes? That would certainly alleviate the guilt from everyone! This affair is not exempt from heated discussions, including the question of biology. What is the current thinking on the role of our genes – is it a matter of nature or nurture?

Those arguing for nurture will cite the fact that human genes couldn't have changed so quickly to account for the soaring obesity rates we are seeing. Hence obesity must be caused more by the environment we live in and our lifestyles. But those on the nature side of the fence, usually biologists, are convinced that the opposite is true, i.e. obesity is caused entirely by vulnerabilities in our genome. Claude Bouchard from the Human Genomics Laboratory at Baton Rouge in the US probably gets closest in the nature vs nurture debate when he says, 'As is generally the case when such diametrically opposed views are upheld, the truth lies somewhere in the middle.' For him it is a truism that the 'obesogenic' environment and behaviour are fuelling the current acceleration in obesity and overweight worldwide. However he also argues that biology has got something to do with it.

Various lines of evidence support the idea that individuals vary in how easily they gain weight, and it seems that genetic variation has much to do with our risk of becoming obese, even more of becoming severely obese. As a matter of fact, all populations appear likely to become obese in favourable circumstances, and this may well be due to the presence of several classes of genotypes that have evolved to keep us in positive energy balance (Bellisari, 2008):

- a 'thrifty' genotype inducing low metabolic rate and insufficient thermogenesis;
- a 'hyperphagic' genotype linked with poor regulation of appetite;
- a 'sedens' genotype leading to a propensity to be physically inactive;
- a 'low lipid oxydation' or 'adipogenesis' genotype pushing to expand the number of fat cells to increase the body's ability to store fat.

This has been incorporated into our genome during the long period of human evolution, helping our ancestors and modern humans to survive when confronted with food scarcity or with high levels of physical activity (such as hunting). This may explain why some unlucky individuals are more prone then others to develop obesity – they have inherited an effective (but lethal!) combination of ancestral energy-conserving genes. Work is still exploratory in this field and may bring new revelations. A range of other biological explanations are being put forward, for example some scientists have suggested that viruses may play a role in cells to expand adipose tissue mass (Rogers et al, 2008). Another recent study, by Kirsty Spalding (2008) of the Karolinska Institute of Sweden reported in the journal *Nature* on the dynamics of fat cell turnover. They showed that regardless of weight, the number of fat cells seems to rise steadily from birth to the early twenties, but then remains constant, even after weight loss, which they suggest is due to tight genetic control.

So what does all this mean? Can obese people do anything to lose weight since they have already accumulated a lot of fat cells while growing up? Yes, fortunately they can still reduce the volume, but not the number, of their fat cells (Shadan, 2008). And what about lean people – do they need to worry about what they eat if they have fewer fat cells? Yes, unfortunately. Fat cells can still store large amounts of fat – even if you don't have that many.

Further evidence for a genetic link came from a large study of twins in the UK (Wardle et al, 2008) that found that BMI was largely down to genes and only a quarter of variation was due to differences in the environment. All these studies appear to point in the same direction: both nature and nurture play a part in who gets fat and who stays slim.

Chapter 7

Go Active!

Athletics at McDonald's? One can't help but smile at the idea. And yet the fast-food chain has been an official sponsor of the Olympic Games since 1976. This idea is not just a publicity stunt; it formed part of a broad global strategy, as the company whose image had taken such a battering, decided to go on the offensive. In 2004 it unveiled a new slogan, tested in the US and various other countries: 'Go Active!™'. What this meant was that, not content with flogging us burgers and fries, McDonald's now proposed to become our fitness coach. The Go Active kit consisted of a plate of salad, a glass of water and… a pedometer to add up all the steps taken in a day, enabling one to monitor one's mileage. The package came complete with a little book of fitness tips. And, to enhance their image still further, the Golden Arches hired bevies of media-genic sports consultants and bought the endorsement of quite a few top athletes, such as the female basket-ball star Yao Ming.

The strategy behind all this was straightforward. Unable to reduce the amount of calories in fast-food meals, lest their true insipidity be revealed, the company threw its considerable weight behind the idea that obesity is not the fault of too much food; it's entirely due to lack of exercise. The food industry seized gratefully upon this theory as it provided a convenient disclaimer for the business as a whole. Basically, you can eat as much as you like, so long as you work out enough.

There's a grain of truth in this, of course. Since obesity is the result of a persistent and long-standing imbalance between the amount of calories a person consumes and the amount they expend, then there are clearly two ways of approaching the problem from a thermodynamic point of view: either too many calories are coming in, or not enough of them are being used up.

Burning off the fat

Let's look at basic physiology for a moment here. The body expends its calories in various ways. It uses some of the energy to keep basic functions going, making sure the heart continues to beat, the brain to work and so on. This is called basal metabolism. Another small amount of calories are dissipated as heat when we digest our meals (we have to use up energy in order to store it) or to protect us against the cold. And the remainder is consumed by our muscles as we move. In the average adult, basal metabolic rate accounts for more than half (between 60 and 70 per cent, depending on the individual degree of sedentarism) of the day's energy expenditure. Digestion itself, and the production of heat if necessary, take up about 10 per cent of the total. Therefore a person's physical activity is responsible for 30 per cent at most of their total energy consumption, that is, barely a third. It's not such a lot after all (admittedly the proportion can climb to 50 per cent in the case of someone engaged in hard manual labour). But since it is practically impossible to vary the base metabolism to any significant degree, physical activity remains the crucial variable in matters of energy expenditure – the only variable we can really play with.

On this point, intuition is confirmed by statistics: it is generally true that overweight people are also those that take least physical exercise. And yet we cannot assume that they are overweight *because* of their inactivity. The causality could just as well work the other way: it might be because they are so heavy, and find physical effort so especially demanding, that they are so inactive. All the same, various studies have found that the decline of physical activity is a prime factor for gaining weight, so that an adult who stops taking exercise during their spare time is liable to put on around 5kg within five years (Rissanen et al, 1991). It is also fairly obvious and not likely to be a coincidence, that no top sportsmen or women are obese; yet they often gain weight after retirement, or when they cease to push themselves as intensively as before. Finally, traditional societies where physical activity is the norm also present very low obesity levels. In short, going active can certainly protect us from both overweight and its consequences. It is too often forgotten that at the time when its health benefits were first described, the famous Cretan diet – like the Mediterranean diet as a whole – was inseparable from regular and sustained physical activity.

The reasons for this are purely physiological. Regular exercise stimulates the body to draw on its stock of fat before that of glucose, so long as the exercise is moderate (in principle, the body uses glucose for vigorous bursts of activity lasting no more than 20 minutes, and only starts drawing on fat reserves after 40 minutes). Hence athletes 'burn off' more fat than others do while making the same physical effort. They can, by the same token, allow themselves to eat more calories than the rest of us.

Good for the figure, great for health

Helping to fight the flab is not the only advantage of physical activity. Performed within sensible limits, exercise is a health benefit in many other ways. According to the WHO, sedentary people are two to four times more at risk of developing type 2 diabetes than those who get a certain amount of exercise, whatever their weight. In other words, a portly person who is always on the go has a better chance of being fit, notwithstanding his or her size, than a slender person who rarely moves.

The official medical recommendation of performing at least half an hour's moderate physical activity every day is actually intended to reduce cardiovascular risks, rather than to make people slimmer, for the simple reason that nobody really knows how much exercise is needed, on average, for an individual to lose weight. But it's undoubtedly a good deal more than half an hour. Some experts reckon that 90 minutes of exercise every day would be the minimum required for someone to appear visibly slimmer. The general consensus is that in the light of our eating habits, one hour might be just enough to avoid gaining a couple of pounds. Is this level of exercise really practicable for the population at large?

The physical activity level

There is a scale to measure an individual's degree of activity, called the 'physical activity level' (PAL). It expresses total daily energy expenditure in units of base-metabolic functioning alone, assuming the body to be at rest. For example, if a person scores a rating of 2, it means that they expend twice the energy as they would if they were to remain completely static.

The WHO's recommended activity level to ensure a stable weight corresponds to a physical activity level of around 1.8, which is by no means easy to achieve. In large cities, sedentary subjects rarely score more than 1.6, and it's quite a challenge for them to exceed this: an adult male weighing 70kg, and seeking to raise his score to 1.7, would have to either practise an energetic sport for 20 minutes a day, or walk for an hour (Ferro-Luzzi and Martino, 1996). Our dogged slimmer would have to keep on walking for over 90 minutes to make the recommended 1.8. And, of course, this effort would come on top of the 24 minutes of 'active leisure' (12 minutes' vigorous exercise plus 12 minutes' walking) required merely to score a paltry 1.6. What a fate, to have to make up for the dangerous comfort of a day at one's desk by gruelling jogs around the park after work!

An armchair society

How ever did we get to this point? It seems that little by little, our societies geared themselves to do away with as much corporal motion as possible. The traditional culprit, of course, is television, and the accusation is justified. It has been established that the more we watch, the greater the risk of girth expansion within just a few years. By and large, children of all social backgrounds who frequently watch TV and play video games tend to be plumper, with higher cholesterol levels, than their more active peers. Unfortunately, watching TV has become the preferred leisure activity for children as it has for adults. Already in 1994, the average British citizen was spending more than 26 hours a week in front of the box, as compared with 13 hours during the 1960s (Office of Population Censuses and Surveys, 1994). And it has become a cliché to say that the typical American child spends more time hunched in front of a screen than at school. These kids are tireless at surfing the Web, networking and gaming; they might also read, or phone their friends. What they hardly do is walk anywhere.

The average number of vehicles per household has risen considerably over the past few decades. It's not unusual for a family to own two or even three cars. Little wonder then that most journeys are made in a car rather than on foot or by bike, no matter how short the distance involved. Many people drive to the corner shop just to pick up some bread.

In the UK, in 1992, children under the age of 14 were already walking 20 per cent less than they did in 1985 (DiGuiseppi et al, 1997). The distances they cycled had likewise shortened by 26 per cent, while they were covering 40 per cent more miles in the back seat of the car. The picture is very similar in France: according to the Centre d'étude sur les réseaux, les transports et l'urbanisme (Networks, Transport and Planning Research Centre, or CERTU), in 1976 more than 80 per cent of children aged five to nine used to walk to school. By 1988 this figure had plunged to around 65 per cent, and by the late 1990s it was under 50 per cent. And when school is out, these children play in the street or in any public space far less than their forerunners did, due to parental anxiety about traffic, strangers and other dangers. In many urban areas, women and the elderly as well as children are reluctant to venture out alone, especially after dark. Another lost opportunity to get active.

In short, everything conspires to make us hoard our calories. At home, toasty central heating saves the body from having to warm itself by burning sugars or fats. At work, machines, computers and other labour-saving devices have nearly abolished movement: the furthest an employee might travel is from their desktop to the printer a few steps away. Only a small minority of people can claim to perform consistently demanding physical tasks.[1]

In public spaces such as shopping centres, convenient lifts and escalators save us time and energy. Doors swing open by themselves, depriving us of even this chance to make a modest effort. Such facilities were installed with the best of intentions, but to the detriment of our health.

Cities themselves have been designed with motor cars rather than pedestrians in mind. Urban play areas are at a premium, and cycle lanes in many countries are still in their infancy. There has been a rapid transition all over the world from neighbourhood-centred towns organized around local shops and businesses, to American-style horizontal built-up areas that, lacking proper centres, can only be negotiated by car, and where the sight of a person on foot attracts curiosity at best, and at worst, suspicion. In the US – but not only there – many streets don't even possess a pavement worthy of the name. And hardly anyone seems to mind. For how can physical exercise be promoted if people don't feel safe in the streets?

In 2008, The National Institute of Health and Clinical Excellence (NICE) in England and Wales tried to address this problem by targeting urban planners in a new report (*Physical Activity and the Environment*) on preventing obesity by modifying the environment (NICE, 2008). They suggest that planning applications for new developments should prioritize ways to keep people active. Transport planners are encouraged to widen pavements and add more, better cycle lanes, as well as to narrow or close access to roads, to slow down traffic and charge motorists for using roads – such as introducing congestion charges, as was initiated in central London in 2003.

Box 7.1 Graffiti is fattening

Does living amid graffiti increase the risk of obesity? Indirectly, yes, as has been found by Anne Ellaway and Sally Macintyre, of the Sociology and Public Health Unit at Glasgow University, and Xavier Bonnefoy, of the WHO's European Centre for Environment and Health in 2005. The three researchers posed the question of whether pleasant neighbourhoods that are nicely maintained and endowed with green spaces really did encourage their inhabitants to get out more, and limit their weight gain through exercise. The survey, conducted in eight European countries, compared the 'quality' of the respondents' surroundings in the immediate radius of their homes (using negative markers such as graffiti, litter or dog mess and positive ones such as green spaces) with their rate of physical activity. It transpired that for an equal social status and income, people who live near green spaces are three times more likely to take above-average amounts of exercise, and 40 per cent less at risk from overweight and obesity. By contrast, those who live in run-down areas where graffiti and other expressions of antisocial behaviour are rife, have a 50 per cent higher chance of being physically inactive, and are 50 per cent more at risk of being overweight or obese.

A question of culture

It's not always easy to bring home to people that they need to be more active, or to take up a sport, especially when people have made a point for decades, if not centuries, of economizing their energy and minimizing their physical efforts as insurance against times of hardship. In developing countries, a woman may have to walk for at least half an hour, sometimes 90 minutes, simply to fetch water. To perform ordinary household chores keeps her busy on her feet for another 90 minutes (WHO, 2000). Even so, the total amount of physical energy expended by the end of the day is not always as great as those figures suggest – because adults in developing countries compensate by switching off whenever they can. When food is scarce, the first human reflex is always to rest. Thus the notion of jumping around during one's spare time goes against the grain, and while the food supply is now much richer in calories, the mindset hasn't really changed. The upshot is that in many Southern countries people continue to exert themselves as little as possible, even though most have more calories to spare nowadays. There's no avoiding an imbalance in such conditions – whereas in Northern countries the problem is less acute because the sports culture is livelier, and the weather is cooler! Faced with the difficulty of promoting participative sports and physical activity in general, societies have fallen back on the search for a pharmaceutical solution. A frantic race is under way to formulate a wonder molecule that could increase the amount of energy we expend with no need to alter our lifestyles. This drug-centred approach to obesity itself raises a host of questions, on both a medical and ethical front.

Notes

1   Alternatively, it is common knowledge than many office workers actually suffer from pathologies such as back pain, tendonitis or osteoarthritis, brought on by inappropriate work conditions, for example bad posture, badly designed chairs or equipment.

Chapter 8
Slimming with Pills

In 1994, researchers discovered that mice that had been genetically modified to be obese grew miraculously svelte when injected with a hormone called leptin, and the fever surrounding the promising molecule grew considerably. The way it worked was simple. Leptin is a hormone secreted by adipose tissue that makes us feel full. It tells the brain that the organism has built up enough reserves and doesn't need to take in more. This led to the bright idea that obese individuals might be suffering from an insufficient output of leptin. Their brains continued to want to stock up, despite the abundance of fat already present in the body. It would suffice to inject them with the leptin they lacked, for their runaway appetites to be reined in.

Alert to the potential goldmine in such a drug, the private biotechnology firm Amgen snapped up the patent at once. But disappointed researchers soon found out that human obesity is a rather more complex affair than the rodent equivalent. Leptin failed to live up to its promise. It did, in some cases, suppress appetite, but further research revealed that fat people do not underproduce this hormone at all – on the contrary, they secrete more of it than thin people. The problem lies in the fact that their brains, loaded with leptin because of the accumulation of body fat, had gradually stopped taking any notice. The brain had become 'resistant' to the molecule. It would be futile to supply it with more.

So long then, to the glittering hopes pinned on leptin. But a drug-based treatment for obesity had at last been floated. And the notion that obesity might be due to biological malfunction, and not necessarily to a lack of willpower, was an idea whose time had come. There was research to be done, enough to keep scientists busy for decades.

The results have not been particularly impressive so far. In most European countries, only two substances have come onto the market. One is Orlistat (trade name Xenical®), also recently approved by the Food and

Drug Administration (FDA) in the US, which blocks the enzymes charged with digesting fats inside the gut. Up to a third of lipid intake is thus prevented from being absorbed before being excreted. A flagship product for its makers, pharmaceutical giant Roche, it is nonetheless pretty ineffectual, causing weight loss of between 3 and 8 per cent (i.e. from 3 to 8kg for a 100kg patient). Undesirable side effects include oily stools, diarrhoea, abdominal cramps and flatulence. in 2009 the EU approved sales of Orlistat over the counter, as is the case in the US, as well as Australia.

The other, Sibutramine, is a new-generation appetite suppressant that acts in the brain as a reuptake inhibitor of two neurotransmitters: serotonin and noradrenaline (norephinephrine). However, doctors are far from unanimous concerning its safety, and it was recently banned from sale in Italy. Among other side effects, it is apt to raise blood pressure and speed up the heart rate, which requires close monitoring. Appetite suppressants have long had an evil reputation, of course; the first generation of such agents, based on amphetamines, were banned in France in 2002 because of their dangerous side effects, some of which appear several years after cessation. Neither this nor the product's limited success as a slimming aid prevented Americans from popping 3 billion units of the most popular version, ephedrine, over the course of 2003. These drugs have a strong placebo effect, which explains in part why they don't work in the long term.

The verdict of the WHO (2000) on weight-loss medication is couched in cautious terms: 'due to the paucity of data, no particular strategy or drug can yet be recommended for routine use', and the WHO (2000) reminds us that 'Weight-management drugs do not cure obesity; when they are discontinued, weight regain occurs'. This is echoed in an in-depth review published in the *British Medical Journal* in December 2007 (Rucker et al, 2007) that concluded that even though anti-obesity drugs modestly reduce weight, the downside is adverse side effects. The authors go on to say that there is a need for longer-term studies to find out whether the benefits of taking drugs really outweigh the risks for health.

The search for a new drug is forging ahead all the same. The biological mechanisms that regulate appetite and weight gain are increasingly well understood, as are the genetic predispositions associated with them. Several routes are being followed for the development of new drugs, some of which should appear on the market any day now: over a score of products are at various stages of development.

Towards a weight vaccine?

In May 2005, a Swiss start-up firm named Cytos Biotechnology announced that it was about to trial an anti-obesity vaccine. The idea here was to prompt the production of antibodies against ghrelin, a small hormone molecule composed of 28 amino acids made by the stomach, which acts, among other things, as a hunger stimulator in the brain. It appears that obese people display abnormally high ghrelin levels after being on a diet, which could well explain the 'yo-yo' effect whereby a person piles the weight back on as soon as they stop denying themselves. The vaccine, if effective, would incite the auto-immune system to destroy ghrelin and therefore lessen the urge to eat. An effective vaccine has been tested in rats but up to now, it has not been approved in humans. There are still questions about the potential side effects of this method. Several alternatives are also under study (for example inhibiting ghrelin secretion or blocking its metabolism to an active form). But as yet, nothing is ready to be used as an effective anti-ghrelin to combat obesity in humans.

Box 8.1 Smaller stomachs

Bariatric surgery, a set of procedures to reduce the volume of the stomach, is the most drastic solution of all to obesity. The WHO regards it as the most effective way to deal with morbid obesity (BMI>40), and to permanently maintain the weight loss achieved. In the US, the number of operations has grown exponentially: a few thousand patients in 1995, 30,000 plus by the year 2000, and this had more than quintupled to over 170,000 cases in 2005! And, all things considered, it may also be the cheapest treatment. This is why the operation is covered by the social security system in France and by the National Health Service in the UK. There are various techniques, such as vertical banded gastroplasty, gastric or bilio-pancreatic bypass, or the installation of a ring to constrict the stomach pouch. A rarity ten years ago, these techniques have massively caught on in France (2000 interventions in 1995, to 16,000 in 2001), prompting the authorities to supervise the whole process more closely, from the selection of patients who most stand to benefit, to the provision of aftercare.

A patient normally loses more than 20kg over the 12 months following the operation, and may regain a few during the 5 to 15 years after that. The SOS study found that ten years on, stable weight loss amounted to between 30 and 40kg, depending on the type of operation involved. The first results of the long-term effects of such interventions in Sweden have shown a reduction in overall mortality (Sjostrom et al, 2007). Even so, the operation is still by no means risk-free. Although the operative mortality is very low (0.16 per cent in Sweden), 'stomach stapling' may result in micronutrient deficiencies, post-op complications or depression. In the US, it has been demonstrated that downstream savings associated with bariatric surgery are likely to offset the initial financial outlay after two to four years (Crémieux et al, 2008).

A recent development in France has involved the stomach in a different way – scientists have implanted a pace-maker into a pig's stomach, which has resulted in a fall in appetite when stimulated, without any side effects. It remains to be seen if this new technique could be applicable to man.

Rest assured that the money men will be studying the prospects for the anti-fat vaccine with great interest. Just as they have been with the drug Rimonabant (trade name Acomplia®), which the pharmaceutical company Sanofi-Aventis has recently had withdrawn from use in the EU (October 2008). Its plans to release the drug on the American market also fell flat as it was rejected over safety concerns about unwanted psychiatric side effects including anxiety, depression and suicide attempts (Christensen et al, 2007) by the US FDA in 2007. Indeed, Acomplia® is not only a rather efficient slimming aid (the molecule depresses appetite) but is also alleged to help quit smoking.[1]

Sanofi-Aventis had hoped its new baby would be a 'blockbuster', that is, a drug with the potential to generate over $1 billion in turnover, but concerns over its safety will no doubt curb its success. In the UK, over 100,000 users of Acomplia® had already been registered by 2008; 720 cases of side effects had been recorded including five deaths, although it is not easy to be 100 per cent sure of the link.

Even so, several pharmaceutical companies have other cannabinoid receptor antagonists (similar action to Acomplia®) under phase 2 or 3 development (i.e. taranabant, surinabant), as it is believed that they may help target abdominal obesity and associated metabolic disturbances. And clinical centres are currently testing the effect of the antidepressant (bupropion) and the anti-epileptic (zonisamide) drugs either alone or in combination.

It should be noted that other potential anti-obesity drugs, including Ecopipam, a selective dopamine D1 antagonist, have been withdrawn at phase 3 of clinical studies because of cases of users developing ideas about suicide, as well as some going on to attempt or even die from suicide.

The Center for Business Intelligence, which monitors the evolution of the pharmaceutical market, runs Obesity Development Drug Summits in the US, dedicated to the latest market developments and scientific breakthroughs in this field. After all, the search for radical solutions to the overweight problem is unlikely to be called off any time soon. Surveys show that people who succeed in losing weight find their lives improve in every respect: social relations become easier, anxious or depressive episodes diminish, while overall wellbeing soars, whatever the weight-loss method used. Some patients who lost 50kg thanks to gastric surgery even stated that they would rather become deaf, dyslexic, diabetic, have a dodgy heart or hideous acne, sooner than return to their former bulk. And when offered a – purely speculative – opportunity to regain the weight in return for $2 million, all without exception said they would refuse the money and keep their figures (Rand and MacGregor, 1991).

Box 8.2 Polypills vs polymeal

In June 2003, the *British Medical Journal* introduced its readers to what it reckoned would prove to be the most significant result for the next 50 years, no less (Wald and Law, 2003)! The fanfare was for a one-a-day pill containing six ingredients – aspirin, a statin, folic acid and three anti-hypertension agents – and claiming to prevent up to 80 per cent of heart attacks and coronaries. This magic 'polypill' was said to control the build-up of 'bad' cholesterol and thin the blood, helping to lower blood pressure. According to the scientists responsible, if everyone took one pill a day who presented risk factors or were over the age of 55, the positive impact upon public health in the Western world would dwarf all measures that had ever been attempted before. The cherry on the cake was that the pill's ingredients were either unpatented or their patents were shortly expiring, so that the cost of such a treatment would be negligible. However, no proof of its efficacy has been demonstrated yet, but clinical trials are currently on the go in India. Was this the miracle pill we'd all been waiting for? In the European press, doctors expressed doubts as to the advisability of putting large numbers of healthy people on medication, when everyone knew that a balanced diet, fewer cigarettes and a spot of regular exercise would produce the same result. One and a half years later, other researchers came up with a new, all-natural blend said to be equally effective while being tastier and even more harmless, called 'polymeal' (Franco et al, 2004). The recipe involved 400g of fruit and veg, 100g of dark chocolate, different amounts of fish, almonds and garlic and… 150ml of wine (a small glass). This too was to be swallowed daily in order to gain, according to the authors, an average 6.6 years of life expectancy, or to postpone heart trouble by about 9 years. Which is it to be: one small pill a day, a partial rethink of dietary habits, or a change, however limited, to one's lifestyle? All the options are on the table.

The patient's denial, the doctor's blindness

The specialists are agreed on one thing. Once obesity has set in, it is very difficult to treat. Strangely, however, obese subjects are seldom regarded by their GPs as suffering from an 'illness' in the conventional sense. According to the US Center for Disease Control, 58 per cent of obese patients have never been offered a word of advice from their primary care doctors on the subject of weight loss – even though these professionals must surely be conscious of the long-term health risks. And yet too often they only sit up and take notice when more specific risk factors, such as hypertension are present. They are on familiar ground with that. It's a clearly identified disease, which they know how to handle.

The WHO, for its part, has criticized the inadequate training of primary care doctors, who often possess only the haziest – or downright misguided – understanding of obesity. They are notably confused about the best way of caring for patients or advising the general public. Again according to the

WHO, medical textbooks place too much emphasis on a handful of genetic or metabolic impairments that may indeed trigger obesity in some cases, but only for a tiny proportion of the hordes afflicted by obesity.

Owing to these gaps in their training, most doctors seem to regard themselves as unqualified to assist their patients in losing weight, and see no point in encouraging them to make lifestyle changes. Treating obesity takes a very long time, and the success rate is low. In the few surveys that exist on the subject, doctors openly admit their reluctance to take on obesity, reports the WHO; some of them, regrettably, accompanied this abdication with the expression of unsympathetic or even offensive prejudices against the obese.

Such indifference from some in the medical profession is a major impediment to timely diagnosis and prevention: patients are frequently unaware that there's anything wrong with them. A survey published by Cancer Research UK in 2005 found that while 65 per cent of British men are technically overweight, only 40 per cent of them recognize the fact. And of these, barely half (20 per cent of the total) had changed their diet and exercise regimes as a result. Many cases of diabetes and countless cardiac or vascular accidents could be avoided if only being overweight was correctly regarded as a full-blown health issue in itself. So, what is the most sensible way to deal with it?

A bitter remedy

The truth is that where obesity and other food-related problems are concerned, the wonder-pill solution may not be 100 per cent safe. Questionable remedies are widely distributed without always taking necessary precautions. A case in point is that of statins. This drug, which has generated record worldwide sales, was originally prescribed to people who had suffered a cardiac episode, because it inhibits the formation of cholesterol. But it proved so effective that in France, as in many EU countries, where doctors are aware of the need for the prevention of cardiovascular disease, it began to be systematically prescribed – in strong doses – to anyone whose cholesterol was a bit too high. This was done without first looking into the patient's eating patterns. Consequently France is now the foremost consumer of statins in Europe, even though it has a lower prevalence of overweight and obesity than other countries such as England or Germany. Some 5 million French citizens are on statins!

To be sure, this molecule cuts cholesterol down. But the treatment is life-long, and known to carry a risk of adverse effects – especially on the liver and kidneys. And it is only truly effective, according to the studies, in approximately a quarter of patients. This means that for every successful prescription, three other people are condemned to taking statins forever more with no real benefits to their health. It would surely have been preferable to look first at prospective patients' lifestyles and diets, and begin by proposing a few improvements in that area. A diet containing greater amounts of non-saturated fatty acids, soluble fibres and phytosterols is likely in many cases to reduce cholesterol to safe levels, with no recourse to drugs. Statins could then be reserved for those cases where lifestyle changes had made no difference. Besides, this molecule may well be the nemesis of cholesterol, but it is helpless against diabetes. Thus it only regulates one limited effect of a complex problem of the whole organism.

The same applies to the various forms of gastric surgery. Although these are supposed to be used only for extreme cases of obesity, and only where no other method has been successful, too often they are demanded by moderately overweight people who merely wish to regain their figure. As though one could fix the problem mechanically, once and for all!

In all fairness, we should remember that such deluded thinking is often encouraged by doctors who have a reflex tendency to medicalize the problem. After all, that's how they make a living. It's also a matter of culture: most doctors are highly knowledgeable about drugs and their effects upon metabolism, but somewhat less informed on nutritional issues, for only recently has nutrition become a specialism of medical training in many EU countries. During the course of their career, moreover, practitioners tend to receive a large part of their information straight from the pharmaceutical industry, whose business it is to sell drugs, not to alter public eating habits.

Attacking obesity with drugs, then, suits all parties in one way or another. Patients hope this magic pill will spare them from having to make hard changes to their lifestyle. The pharma companies make juicy profits from it. And the doctor in the middle lacks the time, the authority and occasionally the competence, to channel the patient toward a more holistic approach. Even if medicines are undeniably useful in tackling a range of specific problems, a healthy diet should also form an essential part of the medical arsenal.

The myth of the ideal weight

There is also some confusion as to the outcome being sought. For a long time, the Holy Grail was something called 'ideal weight', a concept that was pounced on by the media and brandished at the obese in a manner as heavy-handed as it was futile. Since then, most experts have conceded that to aim for an 'ideal' weight is not a useful approach. First, because considerable health benefits may be secured by relatively moderate weight loss, of the order of 5 to 10kg, spread out over a year. Second, because the body deploys a number of mechanisms to avoid losing weight too fast. Thus it may be very hard for the obese to downsize to their 'normal' weight, other than by ruthless repression of the natural urge to eat. Given the 'obesogenic' environment in which we live, this discipline is unlikely to succeed. Clinical trials show that most people cease losing weight after a period of 12–16 weeks' self-denial (equivalent to shedding between 4 and 8kg) and that no more weight is lost after six months, though the benefits of the previous improvement may continue to make themselves felt (National Task Force on the Prevention and Treatment of Obesity, 1996). This level of weight loss is often dismissed as negligible by the patient's family and friends, and yet it has required heroic effort and yielded real health gains. In any case, most diet specialists no longer hold with drastic self-denial – an ordeal for the body that can pave the way for the emergence of disorders such as bulimia or anorexia. And if an over-stringent diet plan should fail – as it almost inevitably will – a drop in self-esteem may ensue, a bout of depression and, ultimately, the return of the weight with a vengeance.

Draconian diets: Best avoided

Despite this evidence, popular magazines continue to trot out all sorts of diets as soon as spring comes along. Any that are prescribed by a genuine doctor and designed so as to take account of a person's behaviour, obesity level, lifestyle and environment are often only really effective in the short term – perfect for those who want to shape up for summer and look good on the beach. The long-term prospects, however, are not quite as rosy. The WHO notes that diets supplying less than 1200kcal per day can certainly slash a person's weight by up to 15 per cent in 10 to 20 weeks, but in the absence of a maintenance programme, most of it will be promptly regained. Part of the reason is that as we grow fatter, our bodies manufacture adipose cells that harbour the excess lipids and release them when the body needs more energy. The trouble is that these cells may empty out, but they don't disappear. They stand ready to fill up again at the first whiff of a bacon butty. This explains why it is so hard to stay slim when one has previously been obese. It makes more sense to level out at 80kg, never letting oneself reach 100, rather than to go up to 90 and then aim for 70. Because once the body has topped 90kg, it will find it much easier to drift up to 100, or even 120.

Studies show that after around a year and a half, more than 90 per cent of dieters will have regained every ounce they lost. And these studies are only concerned with authentic weight-loss programmes, the kind that go beyond tinkering with a couple of basic components, and engage in depth with the entire diet and lifestyle profile of a patient. It goes without saying that the gimmicky formulas served up by the mass-market magazines, typically based on the exclusion of this or that food category, produce – at best – no effect whatsoever. But such fads may equally turn out to be dangerous, if followed carelessly.

Nonetheless, there is a shortage of serious scientific investigation of these issues in the form of controlled clinical trials. Many diets have been put forward, even by doctors, whose long-term efficacy has never been rigorously assessed. As for the few that have been tested under scientific conditions, studies usually carried out on small, homogeneous samples, rather than on large and diverse cross-sections of the population. It would therefore be unwise to assume that one size fits all.

All the same, research does seem to indicate that the most effective slimming diets are the mildest and most gradual. Those that don't demand dramatic sacrifices (a maximum of 500kcal deficit per day, compared to the subject's habitual intake) and are based on the consumption of low-fat foods. It also seems clear that diets only work when coupled with heightened physical activity. It's not enough to reduce energy intake, then; energy expenditure must be increased as well. Group support, as in Weight Watchers-type programmes, appears to be equally beneficial, though such benefits are difficult to quantify. Furthermore, is it necessary to step onto the scales every day? There is insufficient data to decide. More research is currently needed not so much on diets in themselves, but on how best to help slimmers stick with them to the end. What kind of support should be given? How can people be prevented from lapsing? For the moment, there are no ready answers to such questions.

The new wave of functional foods

How about inventing foods that actively assist in slimming or that could mitigate some of the health problems brought on by overweight? Tired of being carped at for making us fat, the food industry spotted a challenge here, an opportunity to rebuild its tarnished reputation while developing fresh sources of revenue. Plant sterols, or phytosterols, have now been incorporated into product lines that claim to correct the imbalances of consumer diets. These substances, which also occur naturally in certain foods, actually do help to lower cholesterol. The Danone group was the first to exploit them, in its Danacol yoghurt. But the competition was quick to jump onto the cholesterol-busting bandwagon, which is expected to ring up sales of over $200 million in 2008. In Britain, for example, Unilever riposted with its Proactiv range. There is even an anti-cholesterol beer on offer.

Mars Inc., manufacturers of the famous chocolate bar, announced in 2005 the creation of a new Nutrition for Health & Well-being business unit, charged with developing sweets, snacks and drinks that might 'serve the nutritional and well-being needs of the consumer' and provide 'real health benefits'. Its inaugural product was CocoaVia, an 80-calorie bar containing vitamins and minerals and preserving the flavonols naturally present in cocoa beans, thanks to a new processing method. CocoaVia was alleged to possess antioxidant properties and an ability to stem cardiovascular disease.

The market for such smart foods grew steadily. In Britain, the annual turnover in this sector leaped fivefold between 1997 and 2003 to reach a peak of some $1.7 billion, according to the Food and Drink Federation, and it has a market worth $7 billion worldwide, as estimated in 2007 at the first international conference in Berlin on slimming ingredients. Numerous ingredients are available with varying levels of supporting scientific evidence, for example green tea polyphenols, conjugated linoleic acid (CLA), resveratrol (polyphenol present in red wine), *Hoodia gordinii* extracts (a cactus-like plant from the Kalahari desert), reuteran (alpha-glucan from a lactobacillus bacteria), proteinase inhibitor II (naturally found in white potatoes), and even calcium from dairy products! The list goes on…

And no doubt more and more will come on the scene – companies are screening thousands of botanical extracts for their beneficial effect on abdominal fat cells and adult stem cells. There are basically five types based on the different ways they act in the body: boosting the burning of fat increasing thermogenesis; inhibiting protein breakdown; suppressing appetite or boosting satiety; blocking fat absorption, and lastly, controlling mood as it also has some influence on food intake. These ingredients may be sold alone as supplements or be incorporated into various foods. And beyond slimming ingredients, another strategy is replacing fats during the manufactured food process, generally with carbohydrates.

But do the trumpeted benefits really exist? First the industry needs to increase the scientific credibility of such products and stop making excessive claims that have damaged their reputation too often. No one disputes that plant sterols help lower cholesterol, but this doesn't mean that they alone can regulate the complex web of an unbalanced diet. To down a Danacol a day may indeed cut down your cholesterol, but it won't stop you from becoming obese or diabetic.

Many manufacturers exaggerate the virtues of such 'health foods', presenting them as a hybrid of food and medicine, and hyping them well beyond what has been scientifically proven. Some have not blushed to coin the term 'medifood'. As a typical overstatement, the early commercials for one yoghurt drink suggested that regular consumption of this would protect a body against infection. The makers were soon ordered to rewrite their copy, for there was no scientific evidence that this product could have any impact whatsoever upon the immune system.

Another dubious marketing wheeze, denounced by paediatricians: food supplements for children. Chocolate-flavoured magnesium concentrate or omega-3 capsules tasting of strawberries offer an easy way out for parents who feel bad about not giving their offspring enough fish, with the added promise of amazing effects on concentration leading to a better performance at school. To many nutritionists, this is plain nonsense. In reality, no amount of supplements will ever make up for an unsound diet. Furthermore, once again, there is not the slightest scientific evidence to confirm the claims of these products.

These days, the buzz from the industry is all about so-called functional foods, that is, foods alleged to assist in a range of physiological functions. This kind of terminology is no more appropriate than the last, since it's the totality of a person's food intake that counts. It has long been understood by public health professionals that an overall nutritional balance must be sought, as opposed to focusing on one element in particular. Breakfast cereals are a case in point. Originally promoted as good sources of fibre, vitamins and minerals, they were gradually adulterated in response to customer cravings for ever greater quantities of sugar and salt, so that now, the vast majority of grain-based products are not really what a healthy child or adolescent should be eating for breakfast. Conversely, producers of more traditional foods have defended their corner with reminders that butter, for instance, is rich in vitamins A, E and D, and it would be a shame to deprive oneself of the goodness. Clearly, there's no sense in praising some products to the skies while condemning others out of hand.

Healthier products at last

Now that consumers have begun to get wise to the potential dangers of eating, a growing number of food companies – eager to rebrand themselves as being good for you – are investing massively in the new 'dietary' or 'low-fat' ranges that occupy ever more shelves in the supermarket. One such convert was Nestlé, which embarked on its makeover in 2003 with the announcement that henceforth all of its output would be themed around health and wellbeing. The new priorities would be wide-ranging, including labelling, newly designed to help shoppers keep tabs on their portion sizes. To deliver on these promises, the company set up various research units endowed with a budget of nearly €1 billion.

Many companies are currently phasing out the use of partially hydrogenated fats (also know as trans fats), which at one time were introduced into practically everything to improve texture and taste, but have lately come under scrutiny for being downright toxic. It has been shown that if more than 2 per cent of energy is ingested from these trans fatty acids, there is an increased risk of cardiovascular disease. The EFSA has since endorsed the idea that it should be no more than 1 per cent. Analyses by the AFSSA have shown that this level is generally achieved in France now, though the concentration is highly variable among products,

Box 8.3 On the trail of nano-foods

Medicine in your lunchbox. Such is the dream pursued by researchers at the bio-nanotechnology department of Wageningen University in the Netherlands. The idea is to insert small quantities of drugs and supplements into everyday foodstuffs by means of tiny capsules that would release their contents on demand, dispatching them to various targets in the body. It remains to be seen whether consumers will be able to stomach this cross between hi-tech, food and medicine. The GM controversy has shown that a technology is apt to be massively rejected if ever it is perceived – rightly or wrongly – as potentially dangerous.

and labelling is not yet obligatory to guide consumers. New processing techniques have also made it possible to leave them out altogether. It's a tall order nonetheless, in that commercial imperatives continue to rule supreme: a branded product must always be yummier than its rivals, regardless of the technical wizardry involved.

Cynical marketeering for some, honest rethink for others – either way, there's an unmistakable ripple and it's spreading, because the public outcry is for real. The industry has an interest in responding. To project an image of concern for public health and wellbeing is more vital than ever for this business, following the lead of the pharmaceutical industry. The 'slimming' foods tick all the boxes: they make the company look good, and they are highly lucrative. No flash in the pan, they're here to stay.

And we shouldn't turn up our noses at them. On the contrary, let's applaud the way corporations are declaring their readiness to be part of the solution to public health problems – so long as we keep shopping. In a country where much of the food supply is industrially produced, there can be no health policy without the participation of the food giants in one form or another. And since everyone has something to gain from this arrangement, who can fault it?

Be that as it may, the new foods, simultaneously skimmed and enriched, and despite their undeniable qualities, follow a familiar logic: eat more (or at least the same amounts) of this and that, and your unbalanced diet will be sorted out. Actually, there's another approach that's never mentioned even though it's the single economically viable solution for all: it can be summed up as 'eat less', especially less of certain products. But who will speak up for it? Not the farmer, intent on selling his produce. Certainly not the manufacturer, nor the retailer. And not the authorities either, anxious as they are to keep unemployment down. Although eating frugally – as Hippocrates recommended long ago – makes sound nutritional sense and enables us to live longer, it still remains economic heresy.

Note

1   Rimonabant blocks CB1 receptors, which are acted upon by the active principle of cannabis (it inhibits the endocannabinoid system).

Chapter 9

Prevention is Better than Cure

The task of turning a roly-poly population into sylphs is proving to be something of a challenge, as we saw in the preceding chapter. In fact the enterprise may be doomed in advance. So ought we make obesity into the lynchpin of future health policy in matters of food and nutrition?

When considering the health of an entire population, as in this instance, there are always two options: should resources and treatment be preferentially targeted at high-risk groups, i.e. persons already displaying high cholesterol and hypertension; or should a universal scheme aimed at modifying our collective environment be adopted, even if it means spreading intervention more thinly? The choice is a crucial one. For while it is true that individuals with excess cholesterol are more likely than others to suffer a heart attack, a greater number of cardiac problems occur among those who currently have what is regarded as a normal cholesterol level. To treat high-risk persons will certainly reduce their chances of a heart attack, and yet by focusing on them, we would eliminate only a minor fraction of the total yearly number of heart attacks. Alternatively, a blanket programme dedicated to lowering – however slightly – the cholesterol rate of the population as a whole, including people who have not yet experienced such problems, would prevent a greater number of cardiac events in statistical terms.

The same calculation applies to obesity itself. Making a concerted effort for the obese to lose weight will be far more costly and inefficient – on a national scale – than adopting measures to prevent the entire population from gaining too much weight in the first place. This does not mean abandoning the obese to their fate, any more than we would those who are at high risk of a heart attack or stroke; it means not allocating all available resources to them. It means trying to close the revolving door whereby the formerly obese, at the end of their treatment, are immediately replaced by the newly obese. In any case, given the swelling proportion of the clinically obese in developed countries, not even the entire available medical resources would be enough to treat them all. The targeted approach is thus, *a fortiori*, more impracticable still in developing countries, where resources are even more limited.

Mission: To nip it in the bud

For these reasons, the goal now pursued by most public health experts is to prevent people from gaining weight, rather than trying to make them lose it. This is certainly the strategy promoted by the International Obesity Task Force and the WHO (2000), which consider that 'concentrating efforts to prevent and manage obesity on people with existing weight problems will do little to prevent the occurrence of new cases of obesity'. The data back up this argument, showing that the proportion of obese subjects mounts steeply as soon as the population's median BMI passes the threshold value of 23.

To lose weight and to avoid gaining it are, of course, two quite distinct processes, which call for equally different strategies. Losing weight is essentially a medical issue, whereas not putting it on is largely a question of environmental conditions. The problem is that our societies have fostered a whole industry – and this includes the majority of women's magazines – that blithely confuses the two, inciting us to slim while encouraging the opposite outcome in every way. It's evidently more attention-grabbing to stage miraculous makeovers than it is to nurture a genuine change of lifestyle habits. In the US we have seen an explosion of weight-loss contests orchestrated by glitzy TV shows, all as spectacular as they are counterproductive: failing to meet the goals they set themselves, the participants invariably end up stouter than they were before. For the real challenge is not to become rake-thin, but to achieve weight stability. And this quest would actually involve far less effort and expense, if only our environment were not relentlessly pulling the other way.

The WHO is sceptical about the chances of prevention. It noted in 2000 that:

Indeed, only two studies have so far been specifically concerned with preventing weight gain in adults, and the short-term results are not such to inspire confidence in the ability to prevent obesity… The fact that obesity rates are rising rapidly and unchecked in most parts of the world casts doubt on whether it is even possible to prevent excessive gains in body weight in the long term.

And yet something must urgently be done, if we do not wish to see the greater part of the world's population become irremediably fat.

Contrary to a widespread assumption, shared by many doctors and scientists, obesity is not necessarily the product of gluttony and/or sloth. Even a marginal discrepancy between energy intake and expenditure can build up over the years, to cause a gradual weight rise without the subject realizing it. Nutritionists reckon that for weight to remain constant, the body must not take in above 0.17 per cent more energy than it burns, over ten years! This does not leave us much room for manoeuvre. Some individuals, blessed by fortunate genes, manage almost effortlessly to maintain an 'ideal' weight all their lives. The human species as a whole, however – like other animals – is much better equipped to fight against want than to cope with excess. The body is pretty good at hoarding energy during the lean spells, but bad at forcing itself to expend more in times of plenty. Once its economy is out of kilter, it is caught in a vicious spiral that induces ever greater mismanagement of the calorie balance sheet. And once the fat has arrived, the body does all it can to make it stay. We need only eat a little more richly every day, or walk just a little less, for the love handles to become a fixture within a few short years.

By the same token, if all the members of a given population ate slightly less energy-dense foods and created more frequent opportunities for exercise, collective rates of overweight and obesity could be kept down. It wouldn't take much it seems, to forestall a rise in the average weight of any population. The problem is to identify the most effective measures for obtaining this outcome.

Fewer calories, with more physical activity

Where food intake is concerned, it's obviously advisable to reduce the amount of calories ingested overall. But there's a large fly in the soup: the most sedentary people – those who go everywhere by car, work at a desk and don't practise a sport – expend very little energy indeed. If they are not to absorb more calories than they can get rid of, they would have to eat very little as well. So little that they would soon be suffering from a serious shortage of vitamins and other micronutrients. It is therefore impossible for such people to lower their energy intake to the level that would match their actual energy needs. In order to keep their weight down, they must instead raise their level of physical activity until it is commensurate with eating the right amounts of nourishing food.

According to the experts, this means maintaining a physical activity level of close to 1.8 (i.e. an energy expenditure of 1.8 times that used by the body when it is completely at rest). In Western society, a moderately active man scores only 1.6 and a sedentary person, 1.4! The latter category cannot therefore avoid putting on weight without compromising their health in other ways. Especially if they are on a low income. Because to feed oneself properly on such minimum calorie levels requires a wide diversity of foods, including plenty of fruit, vegetables and cereals, taking care not always to buy the same ones. This is asking a lot from tight budgets. It's another reason why official obesity prevention campaigns lay such stress on the importance of exercise.

Physical activity is one of the pillars of France's National Nutrition and Health Programme, known as 'the PNNS' (Programme National Nutrition Santé), launched in January 2001. Its recommended 20 minutes' brisk walking a day is not particularly strenuous, and certainly not sufficient to help people reach the ideal activity level of 1.8. But it would get them up to at least 1.6. If everyone set themselves this modest goal – and really persevered with it – the average weight of the French people would shift perceptibly downward. And that's what matters, after all.

So the main objective is to shake the more immobile parts of a given population into performing at least a minimum of activity. Spending as much time as possible on their feet rather than seated, for example: simply by remaining upright for three hours every day, a person can raise their physical activity level from 1.4 to 1.8. Instead of urging people to exhaust themselves in strenuous workouts that they would never be able to keep up, the idea is to coax them into a broad range of low-intensity activities that add up through the day, tipping the balance the right way at the end of it.

Changing individual behaviour: Forget it

The 'recipes' for achieving weight stability, at least in theory, have thus been more or less agreed upon. The next task is to get the bulk of the population to try them out. To this end, divergent methods have been applied. Americans, in keeping with their national culture, think that individuals should take responsibility for managing their own weight. US health authorities have therefore been content to admonish citizens to keep an eye on their waistlines and get more active. It's up to each person to follow this advice or not; should any weight problems develop, they can always seek medical help. A seductive principle, but an inherently hypocritical one. For the obstacles encountered by would-be slimmers are many, and insurmountable in practice given that we live in an environment designed to frustrate every effort in that direction.

The American attitude is a rerun of what was, for a long time, the liberal approach to the fight against smoking: everyone was free to decide whether or not they wished to give up. But in view of the failure of that approach, and the sky-high health costs of tobacco abuse, it became necessary to adopt more powerful methods that were backed up by law. The obesity problem has grown to such alarming proportions in the US that calls for interventionist policies in this area too have multiplied in recent years. The facts are plain: it's no use asking people to change their behaviour without also taking steps to help them do so. The hands-off approach will only work for a very small set of people – those who are exceptionally motivated and well-educated, and flush enough to absorb the financial costs.

The 'free choice' theory is thus unfair by definition, for it's only really a choice for the privileged – precisely those least likely to suffer from overweight! As Dr Anjali Jain put it in the *British Medical Journal* (2005):

> Despite most experts agreeing that the obesity epidemic is due to environmental factors, the research has largely ignored this. It is time to be realistic with individuals about the effectiveness of lifestyle interventions and obesity drugs, and to focus on public health interventions rather than individual treatments, to halt the obesity epidemic.

Scepticism about focusing on individual responsibility, while ignoring the environment was also expressed by England's Secretary of State for Health in July 2008, as he appealed to all sectors, including industry, to work together to prevent obesity, saying that 'We need a national movement that will bring about a fundamental change in the way we live our lives'.

Altering the environment

How can we modify our behaviour in an environment that prevails on us to eat more and at the same time to exercise less? In the multitude of choices we are faced with every day, health considerations, and the perils of obesity in particular, are only a small part of what informs our decisions. To cycle to work, for example, is a good option from the fitness perspective. Alternatively, it could make the journey longer; you have to wear the appropriate gear, and if there's no shower at the office, you may feel the need of one all day. The lack of cycle lanes make it a potentially dangerous means of transport (the health benefits are not guaranteed, in that sense); and there's a major risk of getting your two-wheeled companion stolen if it has to be left in an isolated spot. Finally, it requires a titanic effort to motivate oneself to pedal forth on cold or rainy mornings. In short, everything militates against the bike. Offered a choice between hardship endured for the sake of long-term rewards, and immediate convenience, most of us, quite naturally, opt for the latter – and take the car. The same goes for ready meals, which save us time we regard as precious; we never think about the fat being stored up for the future, and which shows up within months. Or again, it would be highly beneficial to take time off work for an hour at the gym. But what would the boss say to that?

Confronted with the barrage of economic and social pressures, individuals can hardly be expected to resist on their own – not even by grouping with others into consumer associations, whose initiatives are often exemplary, but that are too weak to swim against the tide by themselves. Nor are educational campaigns likely to make much difference since studies show that people equipped with the correct information about food and nutrition are not particularly motivated to act on it. Though most of us have a working notion of what a healthy diet consists of, very few eat in accordance with that knowledge. However, obesity levels continue to rise, at the same time as more and more people are attempting to lose weight.

The problem is even more acute in developing countries, where better-off consumers have no wish to return to the foods of their forebears or to the tiring manual labour they so gladly left behind, both elements being associated with poverty. Hence the need, in rich and poor countries alike, to elaborate a set of social and legislative measures capable of modifying the environment in all milieus. With stricter laws, for example, to compel the food industry to improve the ingredients that go into its products. Restaurants, school kitchens and office canteens could likewise be obliged to answer for the composition of their menus and the size of their portions. Clubs and community groups, for their part, should be invited to treat physical activity as a worthwhile end in itself.

Precisely who would be entitled to intervene and how still needs to be determined. But surely everyone must feel involved in the effort to prevent their society from becoming obese: the individuals concerned and also their families, the health professionals and every other social sector. For example, urban design and planning should now take this into their brief. Due to the ageing of the population we can't get rid of lifts and escalators altogether, but there should always be an alternative staircase; even more importantly, pavements should be maintained in good condition and pedestrian zones expanded in order to encourage strolling, while a network of bike paths might do wonders for cyclist numbers. Countless similar adjustments can be made so as to facilitate the performance of undemanding kinds of exercise that fit in naturally with everyday life. There is an urgent need in American-style housing developments to recreate busy and varied high streets where shoppers can amble from store to store. And the streets must be made safe enough to lure people from the armoured security of their cars. Another crying need is for decent, functional sports facilities, especially in more deprived neighbourhoods.

Public health experts have looked at the impact smoking bans world-wide have had on predicted mortality rates to demonstrate how changing the environment can have an impressive effect. The largest fall in smoking ever seen in England followed the nationwide smoking ban introduced in 2007 in public places, including in the traditional pub. An estimated 400,000 people quit smoking as a result. Initial public opposition to the ban has been short term and demonstrates that radical measures can be taken that make unhealthy behaviours more difficult, and less socially acceptable. This offers hope that changing the 'obesogenic' environment to one that favours healthy eating and physical activity is possible, and can be acceptable to the public in the long term. Obviously smoking is not the same as eating, as we all need to eat, but there are useful lessons to be drawn.

A mosaic of possible measures

There's no doubt that in the interests of efficacy, the combination of measures to be put in place would necessarily vary from one culture to the next. While the causes of obesity appear to be roughly the same everywhere, the weapons to combat it need to take account of the local setting. The one certainty is that regardless of the levels of physical activity that may be attained, no appreciable results will ensue until the intake of calorie-rich products is also reduced. Until, that is, we are prepared to change the basis upon which our societies operate, geared as they are to ever-mounting consumption. Regrettably, the present system incites us to eat more and more, in full knowledge that it is far harder to get us to work off the surplus.

In France, the national nutrition strategy, the PNNS has provided a welcome vehicle for addressing these issues, advancing a set of simple – some would say simplistic – technical solutions that have the virtue of broad public acceptability. Without claiming to monopolize all the answers, this entity represents a first step, helping to show clearly what is at stake, and promoting the first sensible measures, such as the ban on sweet-vending machines in schools and stricter regulation of food adverts aimed at children. The PNNS is also to be commended for highlighting a crucial issue in the debate, namely, the relatively high price of fresh fruit and vegetables, compared to other foods. It has made a strong case for channelling financial assistance to that sector in the form of subsidies or tax relief, even though this would be difficult to accomplish within the framework of current EU legislation.

The idea has been echoed, among others by Adam Drewnowski, Head of the Centre for Public Health Nutrition at the University of Washington in Seattle. 'Simply exhorting people of limited means to eat better is a waste of time', Drewnowski is quoted as saying in *The Lancet* of December 2004 (McCarthy, 2004). Because the sugary, fatty, processed foods such people tend to go for happen to be those best suited to their situation – little money and little time to prepare a decent meal. The arguments of public health, Drewnowski pointed out, are outweighed by economic rationality (Drewnowski and Specter, 2004). He therefore appealed for greater support to be given to programmes distributing free fruit and vegetables in schools and to the elderly. In the long run, he concluded, the solution to obesity lies in raising living standards among the poor, by providing them with better jobs and social services. This deliberately provocative stance reminds us of the limits of the simple solutions advanced by the PNNS in France: there can be no definitive answer to obesity without profound economic change. But it does not mean that economic change will be sufficient by itself.

Probably no single measure will ever be equal to fixing a problem of this magnitude. Only a wide-ranging combination of measures might eventually, if not defeat obesity, then at least bring it under control. Some parents drop their children a few blocks away from the school gates, in order to make them puff a bit. Others are experimenting with the 'walking bus', in which adults take it in turns to walk children all the way from home to school and back. Such actions may sound unambitious, but they could do with being a lot more widespread.

Similar schemes can be launched at the workplace. The Peugeot car manufacturing company has had one in place at its Rennes plant since 2002, after observing that the workers, confined to the assembly line during awkwardly timed shifts, were failing to feed themselves properly and betraying a tendency to nibble, especially on night shifts; they were more overweight than the national average as a result. The employers reacted by inviting each worker to take stock of their nutritional status and then work on it with the help of a dietitian, while the on-site canteen diversified and balanced its menus. In addition, a mobile trolley came round with fresh fruit and dairy products during breaks. To date, the impact upon average worker weight has been slight, but these are the kinds of changes that, when made available to people under controlled conditions such as a factory, can trigger small revolutions.

Many other workplace activities can flourish for an outlay of next to nothing. Rooms can be set aside for playing table tennis and other games. Large companies can hire fitness coaches for the staff. Showers and changing rooms can be installed, for those who wish to cycle or work up a sweat between meetings. The employers have everything to gain from providing such amenities, since overweight and obesity are major underlying causes of lost man-hours, due to anything from diabetes to cardiac problems. The message has got through in Japan, where the office day commonly kicks off with ten minutes of group gymnastics. Over and above the stereotype, this idea makes excellent sense, boosting productivity as well as public health.

Even more modestly, it's good when sitting at a desk to change position as often as possible, to jiggle one's legs as though in an airplane, to get up and move around frequently, to take the stairs instead of the lift, to keep the printer at the end of the corridor rather than close at hand, and so on. It all adds up, at the end of the day, to making a real difference for people who don't otherwise take exercise. But in order for it to really work, employers and employees must be willing to play the game together. They must understand that it's a win–win situation: the employees acquire fitter bodies and sounder minds, the bosses earn more money.

Box 9.1 Say it with low prices

Affordability is more effective than any amount of preaching about public health. This is the gist of the findings by Karen Lock and Martin McKee, of the European Centre on Health of Societies in Transition, based at London's School of Hygiene and Tropical Medicine. In the July 2005 issue of the *British Medical Journal*, the two researchers reported that

> trends in cardiovascular disease in Europe have shown an East–West divide for over 30 years. Rapid declines in the European Union contrast with stagnant or rising trends in Russia and Central and Eastern Europe, with some notable exceptions, such as Poland and the Czech Republic, where rates have fallen since the 1990s. These improvements are attributed primarily to improved nutrition, which can be traced to the economic transition that followed political change in the late 1980s.

So-called unsaturated fats (less harmful to health) and fruit became cheaper and more widely available, causing profound dietary changes and reducing cardiac mortality within a very short time, with no need for costly public health campaigns. Britain itself, according to the researchers, would do well to follow this example, rather than remaining 'focused on medical models of education and behaviour change, even though these have had little impact on rising rates of unhealthy diets and obesity'. Ironically enough, the Polish success story may be imperilled by that country's entrance into the EU, since our CAP subsidizes the production of fats (especially animal fat) at the expense of fruit and vegetable production. We should also note that Poland's impressive reduction of cardiovascular disease in adults has not prevented a rapid rise in childhood obesity there since 1990. This only goes to illustrate the complexity of the public policies that must be implemented, and the need to proceed on multiple fronts at once.

Children first

Given that the objective is prevention of weight gain, rather than weight loss after the event, it's clear that obesity will be more successfully avoided the earlier prevention begins. Therefore, efforts will have to be primarily directed at children and teenagers (which does not of course mean neglecting the plight of adults). And they should begin at birth, if not before. Obesity specialists have come to realize that many cards are dealt *in utero*. If a pregnant woman is diabetic, her child runs a greater risk of overweight in later life. Pregnant women must be carefully monitored, therefore. Paradoxically it has also been observed that children who were undernourished as foetuses or during the first months of life are equally prone to overweight later on – as though their bodies, indelibly branded by memories of privation, were determined never to want for energy again.

Thus it is essential to keep a lid on weight levels from infancy – a stage of whose importance many parents, not to mention paediatricians are not sufficiently aware. Extra work is needed for child obesity to be given the attention it deserves, because overweight is much harder to undo afterwards, once the fat has become established. Prompt treatment of obesity in childhood significantly reduces the risk of that child becoming an obese adult, provided the parents are closely involved. For it's the totality of the child's environment, including diet, leisure activities and TV allowance that will usually need overhauling. Fortunately it's easier for a child or young teen to slim down than for an adult, because their bodies are still growing: even if they conserve a fixed amount of fat, this excess is proportionately diminished as the young person gets taller and his muscles bigger.

Meanwhile it is just as important to make sure the child has access to a diversified diet containing plenty of micronutrients. A plump child must never be subjected to a severe diet: energy should be cut back in moderate fashion, for example by replacing ready-made foods, typically steeped in sugars and fats, with fruit, vegetables and complex carbohydrates such as pasta or rice. And thirst should be quenched with unsweetened drinks, or best of all, with water. This will cut calories while avoiding nutritional deficiency. The danger is that too many severe restrictions could well set off behavioural disorders, such as anorexia or bulimia. Where these occur, the cure will have proved worse than the disease. It is always more effective to substitute what's in the child's plate without him noticing, than to impose a set of prohibitions that will be experienced as punishments. The golden rule is never to let children think that getting thin is the only goal, and never to let them feel like misfits or in any way different from the rest.

Every opportunity for exercise should be welcomed. Competitive games, however, are to be treated with caution, because tubby children, like their adult counterparts, are extra sensitive to ridicule. They are quick to give up on any activity that is hopelessly beyond their abilities and makes them look foolish. To reinforce competitive games in schools may not be the most appropriate course when it comes to encouraging physical activity in the children who need it most. What's more, evidence suggests that weight stability is more effectively maintained by getting the child to limit sedentary activities (TV, video games, etc.) than by increasing physical activities themselves.

The key factor, the most influential of all, is still the family environment. Surveys show that the offspring of non-obese parents find it relatively easier to maintain their own weight. Parental attitudes to food, along with the kind of eating and leisure activities engaged in as a family and the level of support, are primordial. One cannot send the kids to play outside while slumped in front of the telly oneself. It has even been shown that children are better at controlling their weight when at least one of their parents is trying to do the same (Epstein et al, 1994).

Programmes that work

To prove that far-reaching interventions are not impossible, a number of schemes have yielded heartening results. One is Singapore's 'Trim and Fit' programme, introduced in 1992, which is credited with bringing down obesity rates among primary and secondary school students and new university students. The project consisted of a combination of improved school meals, health and nutrition lessons, and enhanced physical activity in the playground. In parallel, teachers and canteen staff were given training and materials on the subject. The consequence was that the numbers of fit children swelled year by year, while obesity levels sank. However, after 15 years of running, the programme was modified due to concerns that it stigmatized overweight children. In its place the Singapore government set up a programme aimed at all children called the 'holistic health framework'.

In the summer of 2008, over 70 mayors and local politicians from throughout the EU met in Brussels to hear about the EPODE (Ensemble, Prévenons l'Obésité des Enfants) project that began in France and now runs in 167 French cities. The EPODE project aims to prevent obesity in children aged 5 to 12 years, by using a community development approach. And the programme has had a certain amount of success: it began in 1992 in two towns in the north of France (Fleurbaix and Laventie) and reports a positive impact on healthy behaviours. Other countries have been so impressed with this approach that they are also piloting it as far afield as Australia and Canada.

Another heartening example is that of North Karelia in Finland, whose inhabitants were suffering from more than their fair share of cardiovascular problems, leading to high premature mortality across the region. A huge cardiac disease prevention programme was launched in 1972, seeking to promote a healthier diet with less reliance on animal fat. Awareness campaigns in the media, community mobilization, public health measures, modifications to the environment, tailored legislation and more: all the stops were pulled out, money no object, in order to orchestrate a mass move toward healthier behaviours. Here too, the results were encouraging, as the average fat content in everyday diets came down from 42 per cent to almost 34 per cent. This was soon translated into a much later onset of cardiovascular disease. The obesity figures, by contrast, remained disappointingly high. But without the programme, they might have been higher still: average BMI tended to stabilize, even though overall physical activity had diminished.

Is it possible to go one better? Unlike tobacco addiction or infectious disease, obesity is linked to a broad range of factors, some of which would seem very hard to eliminate. It seems that in-depth behavioural change cannot be accomplished overnight; it takes time, it's a long haul. Whatever the measures we adopt, the task of overcoming this scourge will be ours for the foreseeable future.

Chapter 10

Some Leads and their Limits

How can the system be brought back to its senses? Through the wallet, in the opinion of some. Economists who are beginning to add up the total costs of obesity have noted a major flaw in the system: those who make a fortune out of the obesity market (chiefly the agri-food industry) are not the same as those who pay the price for it. Common sense suggests that it's only fair for the food industry to stump up for the social costs it begets, a principle that could do wonders for greater corporate responsibility.

I will, if you will

Pressure groups in Britain have been successful in lobbying the government to commission reports on, for example, the total cost of each calorie transported, and to privilege local products as soon as that cost oversteps the line. One such timely report was *Green, Healthy and Fair* (Sustainable Development Commission, 2008), which focused on the government's role in supporting sustainable food in supermarkets, including how they could help prevent obesity. The report describes supermarkets as the 'gate-keepers of the food system' and acknowledges that neither government nor supermarkets alone can resolve obesity. The aptly named report *I Will if You Will*, also by the Sustainable Consumption Commission (2006), emphasizes how different stakeholders have to move forward together. As we saw in Chapter 5, supermarkets have a great deal of power in the world food system and a solution to obesity cannot be achieved without them. They can show commitment by demanding reformulation of products to make them healthier, and ensuring all products on their shelves have simple front-of-pack nutrition labelling. They can also be active in restricting marketing to children in their stores, including getting rid of sweets at the checkout, and some supermarkets have risen to the challenge: the supermarket chain Leclerc is the first in France to ban sweets and chocolate at its checkouts to help prevent childhood obesity. It did so expecting a fall in sales of €5 million, but believed that was a price worth paying for a better image with customers. But not all supermarkets have been so altruistic, as illustrated in a UK National Consumer Council survey in 2008, which found that the proportion of in-store promotions of unhealthy food had more than doubled in two years, prompting criticism that supermarkets were benefiting from the credit squeeze by promoting low-cost processed foods.

On another tack, lawsuits are beginning to loom in the US as obese plaintiffs claim astronomical sums in damages; a sobering development for the food giants, a reminder that they cannot avoid prosecution forever. If consumers were also to band together to insist on different products, the industry would have no choice but to supply the new demand. Except that greater care for health and the environment would inevitably translate into higher prices, which would be bad news for economically vulnerable people in the absence of any economic or fiscal instruments to protect them.

What is certain is that no single approach to solving the obesity epidemic will be enough, and action in a range of settings is needed, from improving school meals to changing the built environment. To achieve this, political will is needed to convince a multitude of different stakeholders to come on board. The WHO has taken the lead by proposing a wide array of ideas to governments on what could be done to prevent obesity (see Box 10.1) and who could do it (see Box 10.2). The buck currently rests with national governments to move this forward.

Box 10.1 Some ideas for what can be done to prevent obesity

Laws and regulations

- Provide nutritional information labelling as proposed in codex guidelines
- Develop controls on food and drink advertising
- Regulate health claims, so as not to mislead the public, for example 'diet', 'light' etc.
- Set up a national coordinating strategy that addresses diet and physical activity
- Develop national dietary physical activity guidelines

Urban design and transportation policies

- Provide incentives to ensure that walking, cycling and other forms of physical activity are accessible and safe
- Increase access to and use of sport and recreation facilities
- Improve public transport (for example frequency and reliability)
- Create pedestrian zones in city centres
- Develop workplace policies that encourage physical activity
- Introduce incentive schemes to encourage use of car parks in conjunction with city public transport (for example park and ride)
- Provide affordable facilities for securing bicycles in cities and public areas
- Install traffic calming measures to increase safety of children walking and playing in the streets
- Modify building design to encourage the use of stairs
- Improve safety by improving street lighting

Economic incentives

- Use taxes to influence the availability, access to and consumption of various foods – could include a 'fat tax' or extending VAT to cover some energy-dense foods
- Introduce subsidies for producers of low-energy foods (especially fruit and vegetables)
- Reduce car tax for those who take public transport to work during the week
- Provide tax breaks for companies that provide exercise and changing facilities for employees
- Provide subsidies to promote access among poor communities to recreational and sporting facilities

Food and catering

- Develop nutrition standards and guidelines for institutional catering services, for example in school and workplaces
- Introduce controls to ensure that catering outlets and vending machines in public institutions sell only healthy foods
- Encourage a reduction in the use of hydrogenated oils and a reduction in the sugar content of beverages and snacks

Food production

- Ensure that European CAP reform is consistent with the protection and promotion of public health
- Introduce incentives to increase or maintain production and distribution of healthier foods
- Encourage use of land in urban areas for growing fruit and vegetables for use by households/families, for example allotments

Promotion of healthier behaviours

- Improve training for healthcare providers (especially primary healthcare) in dietary habits and physical activity, using skill building to change behaviour, taking a life-course approach
- Improve health education for the general public to enable citizens to make informed choices
- Promote applied research, especially in evaluating different policies and interventions
- Use the media to promote positive behaviour
- Educate the public about the main causes of obesity so that stigmatization of the obese is reduced
- Promote exclusive breastfeeding
- Promote the avoidance of added sugars and starches when using infant formula milk
- Incorporate health literacy into adult education

Schools

- Adopt policies that support healthy diets at school and limit the availability of products high in salt, sugar and fats
- Teach media literacy to children
- Provide adequate sport and recreational facilities, including changing and showering areas
- Ensure training in practical food skills for all children
- Provide pupils/students with daily physical education
- Issue contracts for school lunches to local food growers to ensure a market for local healthy food

Source: www.sussex.ac.uk/spru/porgrow; and WHO (2000, 2003, 2004, 2007)

Box 10.2 Some ideas for who could help prevent obesity

Governmental

- Government departments of health (suggested lead)
- Government departments of transport/urban planning, finance, agriculture, food, commerce, environment, social affairs/development, youth, education, family and social care, media and communication, recreation/sports/culture, and parks and forestry

Food production systems

- Food producers, caterers, farmers
- Large and small retailers, supermarkets

Health systems

- Pharmaceutical industry
- Specialist medics, general practitioners, nurses
- Nutritionists, dieticians, health promotion specialists
- Traditional healers, alternative health groups
- Patient groups

Media

- Advertisers
- Newspapers, women's press, TV and radio

Education system

- Schools and colleges
- Pre-school care
- Universities

Workplace

- Trade unions
- Large and small workplace institutions

Non-governmental organizations

- Consumer associations
- Sports groups and associations, walk/cycle groups

- Faith-based organizations
- Low-income associations, marginal group associations
- Parent–teacher associations
- Health promotion organizations
- Childcare organizations

Source: www.sussex.ac.uk/spru/porgrow; and WHO (2000, 2003, 2004)

Taxing junk food

There is growing interest in taxing foods that are harmful to health if consumed in excess. A tax on foods with scant nutritional value can be quite low, as it is in those parts of Canada and the US where the idea has already been adopted. A duty is levied on soft drinks, chocolates, sweets, crisps, savoury snacks and the like, in the form of a fixed sum or as a percentage of retail price.

The surcharge may be small, yet it generates considerable revenue. In Arkansas, the tax on soft drinks (2 per cent on each 12oz can) brings in $40 million a year, and in California, where the duty per can is over 7 per cent, the state pockets $218 million. Across the nation as a whole, the revenue from all such taxes has hit the billion-dollar mark. It has been calculated that a levy of just 1 per cent per can, or per unit of weight of other products of 'scant nutritional value', would be too small to impact negatively on sales but sufficient to yield some $2 billion a year – a tidy sum that could be used to fund educational and preventive projects. The principle, neatly enough, is to force the industry itself to finance the messages that would indirectly serve to deter customers from purchasing its products too often! No wonder the corporations in question are none too keen. Coca-Cola fought back. In the early 1990s it approached the governor of Louisiana with an offer to build a bottling plant in the state, in exchange for a 50 per cent reduction in the tax on soft drinks. Lawmakers went along with it, voting to halve this tax in 1993, the cut to be effective from 1995, and promising to abolish the remaining duty altogether if Coca-Cola agreed to build a second bottling plant, at a cost of $50 million (Jacobson and Brownell, 2000). The contract was signed in 1997 and brought hundreds of jobs with it, putting $3 million into the state coffers. Yet this was small fry compared to the annual $15 million that the state of Louisiana had earned from the very soft drinks tax it had just repealed. A poor deal for the state, then. But what the story illustrates most clearly is the importance of an iron political will to keep such schemes in place, along with the need to marshal enough public support to be able to stand firm against corporate lobbying.

Tax can also be used as a deterrent. High duty slapped on cigarettes, for example, has proved to be a powerful incentive for smokers to give up. In 2001, a poll conducted by Britain's FSA revealed that of all the factors influencing consumer choice, the most decisive was price. Taste and quality were of lesser relevance, while considerations of personal or family health came in last (French, 2004).

In July 2008, the French government of President Sarkozy rejected taxing energy-dense foods because of concerns that such a tax would be unacceptable in France, where food is strongly associated with national identity. Reservations were also that taxes would disproportionally affect the poor, and that the financially important food sector would be hit by any change. Instead the government recommends increasing VAT on snacking, for example on pizzas, quiches, sandwiches. However, the context of soaring food prices makes this politically unsavoury and therefore unlikely. Similar reservations about taxing food by stakeholders had been found throughout Europe in a study of stakeholders in nine EU countries led by Professor Erik Millstone of the University of Sussex. The study, called PorGrow,[1] found that taxation was the least favoured of a range of policy options, meeting resistance by many stakeholder groups. So even if taxation could be effective in changing eating patterns, lobbying would be needed to change widespread objection to fiscal policies. Finland is one country that has shown that national price policy can be effective when coupled with other policies, such as nutrition education and food labelling.

Making healthier food cheaper

Tweaking price levels can act as a powerful lever. It has been demonstrated that cutting the price of low-fat and unsweetened products results in a much higher uptake of these. Tax policy could therefore play the same game, by reducing VAT on, or subsidizing, the items that are to be promoted such as fresh fruit and vegetables. One study conducted in a secondary school indicated that when apples or carrots were displayed in the cafeteria at half price, students bought four times more of them.

At present, the EU's CAP tends to subsidize surpluses of foods with scant nutritional value. A focus of negative press among many, not only nutritionists, the CAP has been blamed for around 9000 deaths a year through heart disease (Lloyd-Williams et al, 2008) and blamed for having a role in the obesity epidemic. However, its effect on health is disputed by some, including Josef Schmidhuber, a senior economist at the FAO, who argues that on the contrary, the CAP may have actually curbed food consumption, including sugar and saturated fats. He suggests that fiscal policies are ineffective in changing food consumption in Europe and other more important drivers are probably 'the overall increase in income, the rise of supermarkets, and changes in food distribution systems, women's participation in the work force, and the growing importance of food consumed outside the home'.[2]

So subsidizing the cost of less energy-dense foods has its limits, not least subsidies such as taxation; it is unpopular with stakeholders from all sides of the fence, from consumer groups to food manufacturers, as was found in the European PorGrow project.

Empowering consumers

Do consumers wield any real power over the food chain, and could they compel the system to make healthier foods available? This is obviously the crux of the matter. In our societies, the customer is king: if a product is not purchased, that product does not exist. Wised-up consumers ought then to be in a strong position to demand and obtain healthier, more nourishing offerings from the industry. But is it realistic to expect them to do so any time soon?

There's already something of a positive backlash in the domain of 'ethical' trade. Fair trade coffee, for instance – which costs a fraction more, in exchange for the assurance of a more equitable return to the grower and the respect of basic environmental standards and social rights – has been gaining ground over the last 30 years, so that it now accounts for more than 10 per cent of total coffee sales in Britain. Is it possible to conceive of a similar groundswell in defence of higher nutritional quality? Could such a movement force the most grossly 'obesogenic' products off the shelves?

For the moment, it would seem that most consumers are firmly hostile to anything that smacks of interference in what they regard as their sacrosanct freedom to eat what they like. It's what they used to say about smoking: hands off my individual freedom to light up! So we're a long way from a broad-based consumer movement coming together to impose restrictions on the products that can be sold in the shops. Unfortunately, these same consumers haven't grasped that given the abundance of mouth-watering stuff, amid contradictory messages and relentless marketing, their freedom of choice is just an illusion. The most noxious products can look, smell and taste just the way contemporary society likes them. The inescapable conclusion is that today's consumer is helpless to exercise any control over the system whatsoever. With unconscious cynicism, shoppers are always asking for rock-bottom prices, and then they are amazed at the impacts of this upon the environment and their own health.

It seems that our politicians are the only actors who could make a real difference, by use of selective taxes, restrictive legislation, clampdowns on supermarket expansion in vulnerable areas and so on. But politicians, in thrall to powerful industrial lobbies, will only act if voters loudly insist that they do; and so far, any such voices remain inaudible. To conquer obesity will thus require a complete new awareness, the re-education of the great mass of consumers, and this seems a distant prospect. But it is not unthinkable in the long run.

Another way of labelling

Implementing mandatory front-of-pack nutrition labelling on foods has been advocated as one possible measure to increase information to consumers to help reduce further escalation of the obesity epidemic. The argument is that consumers cannot be empowered unless they are equipped with the facts that will enable them to make informed choices. Product labels displaying the relevant information can be very helpful here. At present, most labels indicate the number of calories per 100g. That is not particularly enlightening to most people; how many of us bother to look at the numbers, let alone possess the mental arithmetic to convert them – at speed – into some notion of the calorie value of what we're actually going to eat? This is highlighted by Erik Millstone, Professor of Science Policy at the University of Sussex 'Even if you went round a supermarket with a laptop and a set of scales, it would be difficult to make sense of the information provided' (cited in Hyde, 2008). This is why there is currently much talk of introducing simpler and more user-friendly labels, to be standardized all over Europe. Under current European law, products do not need to be labelled unless they make a health claim such as 'low sugar' or 'low fat'. However, this looks set to change. One suggestion is the traffic-light system, which is currently recommended for the UK: green for foods that can be eaten at will, such as salad; orange for those that may thicken the waistline if devoured too often, such as a tasty marbled steak, and red for those that must be consumed sparingly on pain of becoming obese, such as confectionery and other energy-dense or high-fat products. One can already hear the howls of protest from sweet and chocolate manufacturers, and epic battles are being fought in response to the EU proposal for food labelling rules, out for consultation since 2008. At least the traffic-light system would help reinforce the principle that while no foods are inherently good or bad, there are some that are harmless whatever the quantity, and others that are best taken in small doses. The colour coding would bring this home in an instantly comprehensible way.

Can this kind of labelling really fend off obesity? To some extent, perhaps. When we treat ourselves to a bar of chocolate, we know perfectly well that it's not something we should indulge in all the time. A red label would be neither here nor there. Where the colour system becomes useful is in assessing packaged products such as pizzas, table sauces or oven-ready dishes, whose energy contents are not immediately obvious to the consumer. A standardized form of labelling, in which essential information could be reviewed at a glance, might also provide a clearer notion of the product's true nutritional quality. At least for those who possess some knowledge of the subject and could alert the rest of the community through the media or the activity of consumer groups.

Slow Food: The leisurely alternative

As fast foods spread like wildfire and culinary cultures become ever more homogenized, a counteroffensive with a playful touch has emerged, calling itself Slow Food.[3] The international association of that name, founded by Carlo Petrini in Italy in 1986, arose in opposition to the standardization of taste, championing instead the diversity of cuisines linked to unique regional conditions and reviving gastronomic traditions and ancestral techniques. Its emblem – a snail – is a clear image of its philosophy: a leisurely take on life, in contrast to the reigning ideology of frantic productivism, and the ability to feel at home everywhere (like the snail that carries its house on its back). The movement now boasts more than 80,000 members in different countries who share its passion for conviviality, hospitality and deep-rooted *terroir* values. It is structured into small local groups, or 'convivia', that regularly get together over exquisite meals to plan a wine-tasting expedition or a promotional event. Slow Food holds its own conferences and runs its own university, with campuses near Bra and Parma in Italy.

The movement has often been caricatured as an elitist club for a bunch of hedonistic foodies. That critique is not unjustified in some respects, and yet Slow Food must be commended for drawing attention in its own way to a crucial principle: food as enjoyment. If we are ever to overcome obesity and other food-related disorders, eating must remain a pleasurable activity. This is now taken for granted by nutrition experts, but it was not always so. Bring back the joys of the table then, combined with the sociability that instils more structured behaviours. Let's rediscover the sit-down ritual with family or friends, tucking into a good thick stew, a bit on the greasy side perhaps, but relished in good company. Such habits are a better safeguard against obesity than swallowing a quick hamburger in the street. Behind the bluff *bon-viveur* aspects of Slow Food, lie serious questions about our attitude to food and our life choices in general. The movement represents an authentic attempt to explore another way.

However, its inbuilt elitism is a drawback. How can such a movement be extended to include the poor, who, as we have seen, are the main victims of the obesity epidemic? The Slow Food way of life is not likely to be an option for them. In its present form, it is only for people who possess the right amount of money, knowledge and culture. It speaks to those who enjoy enough cultural and material resources to abdicate voluntarily from the dominant model and pursue the adventure wherever it leads. Nevertheless, its basic premise can be rescued to serve as the cornerstone of alternative thinking with the potential to develop into a genuine mass movement some day. After all, when Parmentier set out to popularize potatoes in France, he didn't hand them out among the paupers, he went first to convince the king!

Should we eat less meat?

Until now, governments have responded to the obesity crisis with programmes focused on informing and educating the public, in an attempt to persuade people to adopt healthier eating patterns. But our leaders omit to call into question the other components of the food chain – in particular its methods of production. For it's the whole system that is at fault for making people overweight, including farming.

Among the various elements that are part of this 'obesogenic' agriculture, we have noted the growing importance of meat consumption in the modern world. Experts reckon that the global demand for meat will have doubled by 2020. This will have a staggering impact on farming. Can so many animals be produced – that is to say be fed? No matter which approach we go for (intensive, rationalized or, even better, organic agriculture), there is bound to be massive pressure upon the grain supply, cereals being the basic ingredient of livestock feed. During the 1990s, grain volumes destined for livestock rose by a third in China, and by two-thirds in Indonesia. Will there be enough steaks to go around the 9 billion people expected on this planet by 2050? There's nothing unreasonable about this aspiration on the part of peoples who only wish to live like Westerners do. But it will take a huge toll, on the environment and on their health. Meat is after all the prime source of saturated fats, known to provoke cardiovascular problems, diabetes and cancers. Besides this, as we have seen, most of the recent food scares of the Western world were connected to industrial livestock farming. It is also a notorious fact that in any country whose meat consumption has risen, obesity levels rapidly follow suit. Could this be the key to the simultaneously environmental and nutritional problems besetting the Earth today? Is it all down to the insatiable consumption of meat that overtook the developed world in the second half of the 20th century? We cannot say for sure. But the troubling fact remains that this reckless hunger for animal flesh turns out to be implicated, time after time, in all the problems – environmental and health – that plague our unfit world. We say more on how climate change and obesity are linked in the next chapter.

Producing less, but better

If we are looking for ways to respect both the environment and our health, a return to seasonal consumption might be a good place to start. To splash out on Chilean cherries in winter, or more humbly, apples from New Zealand, is a pleasure that costs the environment dear (see Chapter 4). The revival of locally based networks of growing and consumption would save considerable energy in terms of food miles, while encouraging us to eat in a way that better reflects the surroundings in which we live. It would also protect us, to some extent, from obesity. Care for the environment and war on obesity might thus go hand in hand. The idea is to wean ourselves off the cheap-and-plentiful model, in favour of one that puts quality first; to produce less food, but better food. This is the project ostensibly taken on board by France's INRA, prompting it to restructure its research programmes around issues of food and environmental quality. But 'quality' is a slippery concept, open to interpretation. By and large, the general public understands it to mean 'more tasty', whereas current agronomic scholars view it rather as matching the food on offer with the needs of the food industry. In their minds, to put it frankly, it is not so much a matter of improving the eating experience, as of coming up with the products the industry can most easily offload on us. Taste and nutritional quality will sometimes help it do this, but not always. The same terms are apt to refer to quite different realities depending on who is using them, and we should be on our guard against misunderstandings.

Biotechnology vs bio-ecology

In view of the limits imposed by the environment, contemporary agriculture relies more and more on biotechnology, and the life sciences in general, to increase yields or develop more resistant strains, such as crops that can grow in salty soil. Another exciting line of research involves the production of raw materials by means of micro-organisms. This kind of high-tech agriculture is flavour of the month at present. But a quieter counter-trend is beginning to emerge that also seeks to recruit the sciences, if not the same disciplines: it would use the lessons of ecology, for example, to find ways of minimizing our use of resources and shoring up local ecosystems and customs. The intention is praiseworthy. But it remains to be seen whether this more environment-sensitive agriculture will be equal to feeding the extra billions that will soon be with us. So far it has not demonstrated this capacity – no more than the high-tech solution has. Meanwhile, few peoples on this planet are going to put up with changes to the agriculture and food system in the absence of solid assurances that current performance levels can be maintained. From where we stand at present, it is difficult to predict which one of these two approaches will prove most adequate to meet the challenges of coming decades, particularly in the current climate of soaring food prices.

A global code of practice for advertising junk foods to children

A new code was launched worldwide in 2008 to try and curb the effects of global food and drink marketing to children. The code is backed by the International Obesity Task Force and Consumers International. These organizations are lobbying national governments to adopt the code as part of other initiatives to slow down the worldwide epidemic of obesity and overweight in children. The code focuses on marketing of energy-dense foods that are low in nutrients. No doubt the major obstacle will be getting the food industry to agree what an unhealthy food is, and that's if they accept that such foods exist in the first place!

The international code encourages governments to ban:

- TV and radio adverts between 6 am and 9 pm that promote unhealthy foods;
- use of cartoon characters, celebrities or competitions to market unhealthy foods;
- inclusion of free gifts, toys or items for children to collect in unhealthy foods;
- promotion of unhealthy food in schools;
- marketing of unhealthy foods using new media, such as the internet and text messages.

Some countries have already taken unilateral decisions on controlling food marketing. The Scandinavian countries have been brave enough to introduce controls on advertising junk food; Norway as far back as the

1980s. More recently, Ireland introduced new controls in 2008 on advertising 'junk food' during children's programmes. But this may not go far enough, as children do not only watch TV programmes that target them. Recommendations by Ofcom, the independent regulator for the UK communications industries, are more ambitious, recommending a total ban on advertising energy-dense foods in and around all TV programmes that could appeal to children under 16, broadcast at any time of day or night.

An investigation conducted in 2008 by Consumer International has revealed the extensive lengths food and drink companies will go to when marketing unhealthy products to children in Asia. The report, *The Junk Food Trap*, reveals how major players such as Coca-Cola, KFC, McDonald's, PepsiCo and Nestlé use a persuasive range of marketing techniques to influence the food choices made by children and adolescents: from the direct attraction of celebrity and cartoon character endorsements to internet promotions. The concerns are obvious – marketing unhealthy food is coaxing children into eating more calories, not good news in the context of soaring rates of childhood overweight and obesity in the region. The scale of marketing is shocking and the study finds that aggressive marketing techniques are being used in poorer countries that the same companies have agreed to curb in wealthier nations. The report cites the case of Malaysia, where KFC have set up the 'Chicky Club' to promote their children's menu, which is now the most popular kids' club in the country, with over 50,000 members. To date, Consumer International found the response from governments and the food industry falls far short of what is needed.

Notes

1   www.sussex.ac.uk/spru/porgrow
2   www.fao.org/es/esd/Montreal-JS.pdf
3   www.slowfood.com

Chapter 11

Obesity and Climate Change: An Odd Couple?

Evidence for an unexpected relationship

Obesity and climate change are both crucial issues of global concern. At first glance it may seem odd to link them together and to suggest that obesity is contributing to climate change. Can what we choose to eat really have an impact on global warming? Around 33 per cent of the world's adult population (1.3 billion people) is now obese or overweight (Kelly et al, 2008). In addition, carbon emissions are high and have increased from 250ppm (50 years ago) to 380ppm in 2007 (Egger, 2008). Some have suggested that it is no coincidence that countries with higher obesity rates tend to have higher carbon emissions, such as the US. Recent contributions to mainstream scholarly journals, such as *The Lancet, British Medical Journal, New Scientist* and *Obesity Reviews*, all highlight the interrelation between obesity and climate change, emphasizing how their causes and policy solutions are linked, putting the issue on the public agenda. The UK government commissioned the 'Foresight report' (Department of Health, 2007) with a view to answer the question of how a sustainable response to obesity could be delivered, and Alan Johnson, the UK Secretary of State for Health, warned that the obesity crisis is as serious as climate change for Britons. To support the UK government's ambition to be the first major country to reverse the growing tide of obesity, a strategy document followed in 2008 *Healthy Weight, Healthy Lives: A Cross-government Strategy for England.* This document, as well as the Foresight report, make clear links between both the causes and the solutions to both obesity and climate change, stating that the 'causes of excess weight are similar to climate change in their complexity' (Department of Health, 2008). The financial impact of obesity and overweight is now starting to be felt and the cost to the UK economy alone is an estimated £10 billion a year, which is projected to increase fivefold in the next 40 years due to escalating obesity rates (female obesity has almost tripled and male obesity quadrupled in the last quarter of a century). Many of the costs from increasing obesity worldwide will be carbon intensive, such as increased reliance on medical services and use of drugs for 'treating' obesity, as well as managing its health consequences: cardiovascular disease, type 2 diabetes and some cancers, to name a few.

Box 11.1 Greenhouse gases

What gases are involved? Since the Industrial Revolution, human activity has resulted in an increase in greenhouse gas emissions, i.e. methane (CH_4), carbon dioxide (CO_2) and nitrous oxide (N_2O), which are the main contributors to a rise of the global temperature of 0.4°C since the 1970s. Twenty-two per cent of global greenhouse gases come from agriculture (McMichael et al, 2007) and livestock production accounts for about 80 per cent of this. CH_4 and N_2O are closely related to livestock production and are a greater by-product of this sector than is CO_2. The FAO (2006) estimates that the livestock industry generates 9 per cent of CO_2 from human-related activities, 65 per cent of N_2O and 37 per cent of human-induced CH_4, and is mainly produced by the digestive system of ruminants (enteric fermentation). Although low-income countries produce only 20 per cent of CO_2 emissions, they produce more than half of N_2O and nearly two-thirds of CH_4 emissions.

Similar causes?

As we have seen earlier, the bottom line is that excess weight is a consequence of an imbalance between energy intake and energy expenditure. An increasing consumption of food, especially energy-dense processed foods, accompanied by a reduction in physical activity are key factors in the development of both obesity and climate change. The complex relationship between obesity and greenhouse gas emissions is shown in Figure 11.1. Food production makes a significant contribution to carbon emissions; as for example, each person in the UK is responsible for producing 10 tonnes of carbon emissions every year – which is equally split between food production, distribution and retailing; energy used in buildings; transport/travel; consumption of goods and services other than food (Griffiths et al, 2008).

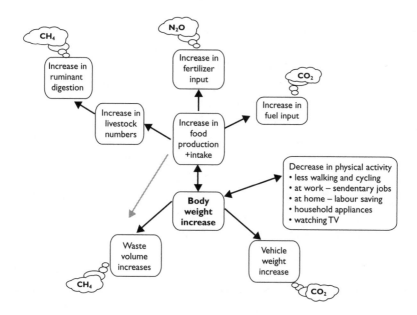

Figure 11.1 How obesity and greenhouse gas emissions are linked

Source: adapted from Michaelowa and Dransfeld (2006)

Drive less, cycle more, save the planet!

The three key drivers of how a lack of physical activity contributes to climate change are first the rise in car use, which reduces the amount of regular exercise we are taking, particularly on our way to and from work, but car use also has the effect of increasing carbon emissions. Second, while at work we are more sedentary, often sitting down for most of the day; service and commercial sector sedentary jobs have gradually replaced those in agriculture in developed countries, which is believed to reduce daily energy expenditure by as much as 1000kcals (Egger, 2008). And when we finally return home from work, there are a host of labour-saving household appliances, such as the washing machine, dishwasher and vacuum cleaner, that reduce the effort we need to make, and therefore the number of calories burned. Such helpful appliances also have the unfortunate effect of increasing carbon emissions.

A recent report from the Institute for European Environmental Policy, based in London (Davis et al, 2007), reviewed the evidence for linking car use to climate change and obesity. The authors note that it is only since the end of the Second World War that private cars have replaced cycling and walking as the main means of transport, particularly for trips to work. They report that many short journeys (of under 1 mile – about 1.5km) are now made by car. For example, it is estimated that around 40 per cent of car journeys in the UK are under 2 miles, which could be walked in less than 30 minutes. The report goes on to suggest that the dramatic reduction in physical activity that followed widespread car ownership is the driving force behind the obesity epidemic. Other scientists (Frank et al, 2004) have estimated that each additional hour spent in a car every day is associated with a 6 per cent increased risk of becoming obese. So the relationship appears unquestionable.

Davis et al (2007) report that main car drivers walk half the distance and for half the time of adults who don't own cars, leading to almost 1 hour less walking every week, which the authors suggest could result in a hefty weight gain of 14kg (2 stones) over a decade – enough to tip most individuals into overweight or obesity. This again reinforces the scientific literature – that it is small amounts of difference in the energy balance equation that lead to overweight and obesity gradually gaining ground slowly over time, and therefore casting doubt on popular stereotypes of gluttonous overeaters. Passenger cars now account for more than 13 per cent of CO_2 emissions in the UK and therefore make a significant contribution to global warming. Reverting back to the walking patterns of the mid-1970s would result in a reversal of the obesity epidemic and a vast reduction in CO_2 emissions. Like several others, Davis et al (2007) suggest that redesigning the built environment to make it more favourable for walking and cycling would be a solution to both obesity and climate change.

An even less obvious relationship between obesity and climate change has been proposed by the influential Hamburg Institute for International

Economics (Michaelowa and Dransfeld, 2006). They suggest that heavier individuals use more fuel when using transport, such as cars, planes or trains. They calculated that an additional 3.4 million tons of CO_2 emissions are produced for an average extra 5kg of body weight of an EU citizen. They also estimate that if each individual in developed countries watched one hour less of TV everyday, then CO_2 emissions would be reduced by 25 million tons. And as we have seen earlier, watching television is more likely to make us pile on the pounds, so such a change would also most probably have a positive impact on obesity. The same logic could also be applied to using computers.

We (and the planet) are what we eat

The food chain contributes an estimated one-fifth of total UK greenhouse gases emissions and is a major source of waste (Sustainable Development Commission, 2008), but why does food production matter so much? The world population was 2.5 billion in 1953, is currently around 6.7 billion and is projected to rise to 9 billion by 2050, which will require a massive increase in global food production, as well as a change in how food is distributed and what is eaten.

The increasing demand for convenience food leads to increased CO_2 emissions because of the production and processing required, as well as the fact that packaging is carbon intensive, as many prepared foods use plastic packaging that is oil dependent (Stern, 2006). The rise in demand for convenience foods has contributed to a diet that is more energy dense, and therefore more obesity promoting, than ever before (Cordain et al, 2005). European policy analysts (Michaelowa and Dransfeld, 2006) estimate that reducing consumption of energy-dense foods back to intakes that Europe had in the 1990s would make enormous cuts in CO_2 emissions (a reduction of over 100 million tons of CO_2). These types of foods have often travelled many food miles and are therefore carbon intensive (also as a result of intensive processing). Increased consumption of energy-dense convenience foods then increases the risk of developing obesity (especially from eating highly palatable, high-fat foods). By contrast, cooking food from basic ingredients is likely to be less carbon intensive.

Another way that the diets of obese people have been linked to climate change is the fact that they need to eat more calories to meet their basic needs to maintain their body weight, i.e. they will need to eat larger portions – about 40 per cent more calories than their leaner peers, according to Professor Ian Roberts (2007), writing in the *New Scientist*. Eating large quantities means that as well as having higher carbon consumption than a leaner person, the obese also produce more organic waste, including methane production when the waste decomposes!

Concern about the rapid worldwide growth in meat consumption on climate change and health was voiced recently in *The Lancet* by the Australian Professor, Tony McMichael and colleagues (2007). They suggest that intake of red meat should be limited, first because it increases the chance of getting certain cancers, especially bowel (colorectal) cancer, and second because eating red meat is associated with heart disease, because of its fat content, and this also links with obesity. However, it is the consumption of the products of the dairy industry that are especially energy dense; milk and cheese in particular push up calorie intake, and therefore increase the likelihood of gaining weight.

The average world meat consumption is 100g/day per person, but this average figure masks the huge diversity of intakes. For example, in developing countries the average daily meat consumption is 47g/day, while it is a whopping fivefold higher in developed countries, at an average of 224g/day (McMichael et al, 2007). Of great concern, as voiced earlier, is that meat consumption is rising, especially in countries where consumption was previously low but that are in rapid economic and nutritional transition such as those in South and East Asia. For example China's consumption of meat has doubled over the last decade (see Figure 11.2). Based on current trends, by 2050, global meat production and milk output are estimated to have doubled from levels in 1999–2001. This has obvious implications for greenhouse gas emissions, as well as for obesity. Professor McMichael goes on to suggest that a global solution to reduce the impact of red meat consumption on the climate and on health problems, including obesity, would be a lowering of consumption in high-income countries per head to 90g/day (of which no more than 50g/day should come from red meat from ruminant animals), which would allow lower-income countries to converge towards this level. This will require an unprecedented shift in the eating habits of most individuals.

However, livestock are not all bad news – for obesity or the environment. Livestock production helps maintain biodiversity, landscape

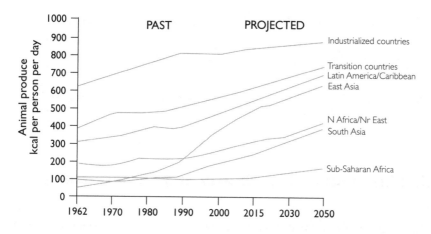

Figure 11.2 How worldwide consumption of animal produce is changing

Source: McMichael et al (2007)

and soil quality. For many poor farmers in low-income countries, livestock are also a source of renewable energy and of organic fertilizer for crops. Red meat consumption protects against iron deficiency, the most prevalent micronutrient deficiency in the world affecting over 1 billion people, particularly pre-school children and reproductive-aged women; it also has a serious impact on school children and working men. If untreated, it can lead to anaemia, which can have severe consequences. The British Meat and Livestock Commission suggest that eating meat can be made more sustainable by choosing British meat, which would have less transportation carbon costs, and by changing what cows are fed to reduce methane production. The jury is out at present on whether eating organic meat is a less carbon-intensive option.

Box 11.2 Can breastfeeding really prevent obesity and reduce carbon emissions?

Recently the protective effect of breastfeeding for both obesity and the climate was highlighted in the *British Medical Journal* (Myr, 2008). It is widely accepted that exclusive breastfeeding for the first six months (as recommended by the WHO) could reduce the number of overweight children. It has also been suggested that breastfeeding means that there is less need for dairy cows to produce milk for infant formulas, which requires diversion of foodstuff to feed the cows. Additionally, bottle-feeding uses materials and energy to modify, package, market and distribute modified cow's milk that all increase carbon emissions.

Similar solutions to obesity and climate change

Some of the policies mentioned previously in Chapters 9 and 10 to prevent obesity could also reduce greenhouse gas emissions. Therefore messages are consistent. Over half the world's population are city dwellers; therefore changes in urban design are fundamental to making physical activity easier and the norm. This would benefit both carbon emissions and individuals' body weight. Some researchers (Woodcock et al, 2007) have suggested that radical action is needed to restrict car travel combined with measures to encourage walking/cycling. A low-carbon transport system involving walking/cycling will help to reduce obesity and it should be a priority for national and local governments to provide safe cycle lanes, footpaths and extensive public transport routes.

Ensuring sustainable catering and food procurement policies is one possible approach, so that local foods are sourced wherever possible and particularly for basic foods produced with minimal processing. Healthy, less calorific foods are therefore more sustainable for the environment, as they are less carbon intensive.

Strategies for managing climate change including personal carbon trading have been advocated as a means to reduce obesity in populations by increasing energy expenditure and decreasing energy-dense food intake, as well as cutting carbon emissions contributing to climate change (Egger, 2008). This would involve a carbon bank attributing carbon units to every country, and each individual would have a set level of units that could be redeemed when buying a non-renewable fuel. There are also suggestions that this could be adopted by the food industry, which would make high-calorie foods more expensive. Most obesity experts would accept that no single approach can tip energy balance sufficiently to influence obesity and carbon emissions, therefore a whole array of strategies is needed.

The potential of supermarkets to influence greenhouse gas emissions was recognized in a report commissioned by the UK government (Sustainable Development Commission, 2008), which states 'as gatekeepers of the food system, supermarkets are in a powerful position to create a greener, healthier fairer food system through their influence on supply chains, consumer behaviour and their own operations'. The report goes on to suggest that existing conflicts need to be resolved between how diets can be both healthy and sustainable (including sustainably sourced fish, meat and dairy). A policy that encourages supermarkets to demand reformulated products and shift marketing to healthier foods will help obesity, but will also direct consumers away from more carbon-intensive food products.

Educating the public to change their attitudes to both obesity and behaving in a more sustainable manner will not be enough. Attitudes are not necessarily a driver of behaviour and it has been suggested that changes in attitude are more powerful when they result from a change of behaviour (Egger, 2008), such as making some behaviours taboo. So sometimes radical changes in the environment need to be imposed (such as banning smoking in public places) to shift society's attitudes. Egger (2008) sums this up by saying, 'regulate and legislate where you can; educate and motivate where you can't'.

Epilogue
At the Crossroads

Is the whole world fated to become obese? Are we to look forward to societies in which everyone, give or take the odd exception, will be overweight? This day is not very far off in the US already, while countries such as the UK or Greece look set to be next. France herself is moving, slowly but surely, down the same road. It's a trend that threatens dire consequences for our health and that of the planet.

But perhaps this fate can be avoided. There's still room to hope that as further reports begin to hit the headlines, each more shocking than the last, we may be galvanized into action. The recent clampdown on smoking is a good example of the way the authorities, with the support of public opinion behind them, are capable of engineering lasting behavioural modifications. They did not hesitate to clobber smokers financially while standing up to corporate interests.

Can the same revolution be fought in eating behaviours? Two political science scholars, Rogan Kersh and James Morone (2002), have shown that when societies find themselves confronting a problem of critical proportions, they will mobilize en masse and support the necessary political remedies, provided three conditions are in place. First, the population at large must perceive that a problem exists. This was the case with tobacco, and it is manifestly the case with obesity. Second, there must have been a steady build-up of scientific evidence detailing the harmful effects of the emergency, and assigning the responsibilities for it; and these scientific data must have been debated, acknowledged and accepted by society. This was the case with the role of tobacco in the development of certain cancers, and it is becoming true for obesity as well: practically everyone is now informed to some degree about the damage done to our bodies by too much fat. Third, and perhaps most importantly, there must be innocent victims. People who have been wronged, who move us to pity and outrage. Well, who is best placed to make us feel this way? Who is the quintessential victim, if not the child?

When obesity was perceived as primarily an adult issue, the popular attitude to what was seen as weak-willed gluttony remained unsympathetic. But the steep rise of child obesity and the proliferation of chronic illnesses such as diabetes, with all that this means for the child's future, was a different matter. Nobody could be so callous as to blame 'the kids'. Blame them for what, anyway? For eating what they were given? And what about society's duty to protect them? The disastrous inroads made by obesity into the world of childhood are forcing the community to sit up, reconsider its preconceptions and consent – at long last – to stern measures being taken. Just as it did a few years ago, to slow the rise of tobacco addiction across the industrialized world.

This time, however, the issue is obviously a great deal more complex. Telling people to stop smoking is easy enough, but it would be nonsense to tell them to stop eating. Tobacco puts nothing but toxins into the body, whereas food – and that includes fat – is indispensable for the proper functioning of the organism. There's no room for simplistic slogans here; it will all be a question of nuance. The measures adopted must draw a careful distinction between reasonable and excessive consumption, while taking account of each individual's whole environment. In short, it will be necessary to proceed with subtlety, at the same time throwing the net wide.

A further difficulty can be foreseen: whereas the tobacco industry was always confined to a small number of companies, the food industry has grown to be an immense hive of economic activity that employs, in Europe alone, hundreds of thousands of people. All of them have a stake in getting people to eat more. Can we allow ourselves to sacrifice whole sectors of the economy? Some tough and painful choices will have to be made. The politicians will probably be obliged to hold off until our system of production and consumption has become so thoroughly untenable, so absurdly wasteful and its health costs such an unsustainable burden on the community, that the urgent need to do something about it will be glaringly obvious to all.

Poorer countries, on the verge of industrialization, would be wise to learn from our mistakes. But can they really accomplish their economic transition while sidestepping the dangers of the food transition that goes with it? There are reasons to fear, on the contrary, that the effects of the transition will jeopardize their health even more severely than it has ours. In the first place, because for centuries such populations depended for their survival on the ability to store any fats that came their way, and have thus evolved a genetic predisposition to 'stock up'. Second, because undernourished embryos or babies are more likely to grow into overweight adults, susceptible to chronic diseases such as diabetes. The war on obesity is therefore intimately bound up with the continuing war on hunger and malnutrition, especially among pregnant women and children. Of course, there's a chance that developing countries may be forewarned by the drastic measures our own societies will have taken. But only if they in turn find the strength to withstand the pressure from financial interests, which will portray any regulation as a menace to the economic growth that is – as nobody disputes – crucial. Carried away by the novel delights of consumption, can newly prosperous peoples be made to see the sinister side of our bloated societies before it's too late? Probably not. It's a safe bet that they, too, will shortly be getting acquainted with the downside of affluence.

The great question is not so much to know whether people will respond or not: after all, the history of our species has repeatedly demonstrated human capacity to adapt and survive through the most appalling ordeals. The issue is whether they will come to their senses before a major emergency sets in. To put it bluntly, how many dead diabetics will it take, how many crippled or blind teenagers, how many heart attacks in the prime of life, how many ravaged ecosystems and how much poisoned waste, before our societies finally decide to change their ways? And if the scale of the human tragedy fails to shake us out of our complacency, the economic costs of it undoubtedly will.

The mad cow crisis was an instructive rehearsal of the process. Only when faced with the possibility of hundreds of thousands of deaths (a realistic theory at one stage), and the collapse of an entire agri-food sector, were radical measures put in place. Today's consumers are caught in a hopeless contradiction: they claim in all good faith to want quality foodstuffs, safe to eat and produced in a way that respects the environment and wellbeing, and then they go and buy the cheapest products on the market. The most committed might buy 'organic' for a special meal at the weekend, but stick with 'non-organic' during the week. Is this really so shocking? Rather than huffing and puffing about it, we should finally understand that none of us are immune to the pressures and constraints, both social and economic, of the environment in which we live. Which harassed wage-earner has the time to cook a decent meal from scratch, or to exercise for an hour a day? What will induce low-income households to spend more money on supposedly healthier products? Who will make it worth the farmer's while to farm in a more sustainable way? How can food manufacturers maintain their profit margins on the back of more soundly balanced, less tasty products? To curb obesity will require profound alterations to the way we live, enforced by stringent rules and regulations, and the enshrinement of policies dedicated to bolstering nutrition and health. These should include, among others: commitment to the principle of higher quality, more expensive products, with assistance for families on low budgets; restrictions on the sale of potentially harmful foods and subsidies for those that are beneficial; strict controls on advertisements for excessively fatty, sugary and/or salty products; improved consumer information enabling shoppers to exercise genuine freedom of choice; and the redesign of urban centres so as to encourage physical activities of every kind. In its July 2005 issue, Fiona Godlee of the *British Medical Journal* expressed the following warning: 'We can tell people what the healthy choices are, but unless we make it easy for them to make those choices, through sensible political and economic reform, we may cause only greater inequity.' For all of us to rethink the way we live will be no easy task. But do we have any other choice?

Bibliography

Chapter 1

Charles, M. A., Eschwège, E. and Basdevant, A. (2008) 'Monitoring the obesity epidemic in France: The Obepi Surveys 1997–2006', *Obesity*, vol 16, no 9, pp2182–2186

European Association for the Study of Obesity: www.easoobesity.org

Guiliano, M. (2004) *French Women Don't Get Fat*, New York: Knopf

Guilbert, P. and Perrin-Escalon, H. (2004) *Baromètre Santé-Nutrition 2002*, Paris: INPES (Institut National de Prévention et d'Education pour la Santé)

International Association for the Study of Obesity: www.iaso.org

International Obesity Taskforce: www.iotf.org

James, W. P. T. (2008) 'The epidemiology of obesity: The size of the problem', *Journal of Internal Medicine*, vol 263, no 4, pp336–352

Kelly, T., Yang, W., Chen C. S., Reynolds, K. and He, J. (2008) 'Global burden of obesity in 2005 and projections to 2030', *International Journal of Obesity*, vol 32, no 9, pp1431–1437

Lobstein, T. (2008) 'Obesity in children', *British Medical Journal*, vol 337, p669

Nestle, M. (2007) *Food Politics: How the Food Industry Influences Nutrition and Health*, University of California Press, Berkeley, CA

WHO (World Health Organization) 'Global Database on Body Mass Index': www.who.int/bmi/index.jsp

WHO (2000) *Obesity: Preventing and Managing the Global Epidemic*, Report of a WHO Consultation, WHO Technical Report Series 894, Geneva: WHO

Chapter 2

Canoy, D., Wareham, N., Luben, R., Welch, A., Bingham, S., Day, N. and Khaw, K. T. (2005) 'Cigarette smoking and fat distribution in 21,828 British men and women: A population-based study', *Obesity Research*, vol 3, pp1466–1475

Colchester, A. C. and Colchester, N. T. (2005) 'The origin of bovine spongiform encephalopathy: The human prion disease hypothesis', *The Lancet*, vol 366, no 9488, pp856–861

Gortmaker, S. L., Must, A., Perrin, J. M., Sobol, A. M. and Dietz, W. H. (1993) 'Social and economic consequences of overweight in adolescence and young adulthood', *New England Journal of Medicine*, vol 329, no 14, pp1008–1012

Hitchen, L. (2007) 'Deaths related to obesity take over from suicide as leading cause of maternal death', *British Medical Journal*, vol 335, p1175

Narbro, K., Agren, G., Jonsson, E., Larsson, B., Näslund, I., Wedel, H. and Sjöström, L. (1999) 'Sick leave and disability pension before and after treatment for obesity: A report from the Swedish Obese Subjects (SOS) study', *International Journal of Obesity & Related Metabolic Disorders*, vol 23, no 6, pp619–624

Poulain, J. P. (2002) *Sociologies de l'alimentation*, Paris: PUF editions

Staffieri, J. R. (1967) 'A study of social stereotype of body image in children', *Journal of Personality and Social Psychology*, vol 7, no 1, pp101–104

WHO (2000) *Obesity: Preventing and Managing the Global Epidemic*, Report of a WHO Consultation, WHO Technical Report Series 894, Geneva: WHO

WHO/FAO (Food and Agriculture Organization) (2003) *Diet, Nutrition and the Prevention of Chronic Diseases*, Report of a WHO/FAO Expert Consultation, WHO Technical Report Series 916, Geneva: WHO

Williamson, D. F., Madans, J., Anda, R. F., Kleinman, J. C., Giovino, G. A. and Byers, T. (1991) 'Smoking cessation and severity of weight gain in a national cohort', *New England Journal of Medicine*, vol 324, no 11, pp739–745

Chapter 3

Barker, D. J. P. (ed) (1992) *Fetal and Infant Origins of Adult Disease*, London: British Medical Journal Publisher Group

de Souza Valente da Silva, L., Valeria da Veiga, G. and Ramalho, R. A. (2007) 'Association of serum concentrations of retinol and carotenoids with overweight in children and adolescents', *Nutrition*, vol 23, pp392–397

Eisinger, P. K. (1998) *Towards an End to Hunger in America*, Washington DC: Brookings Institute Press

Fraser, B. (2005) 'Latin America's urbanisation is boosting obesity', *The Lancet*, vol 365, no 9476, pp1995–1996

Garrett, J. L. and Ruel, M. T. (2003) 'Stunted child-overweight mother pairs: An emerging policy concern?', FCND discussion paper no148, Washington DC: IFPRI

Gibney, M. J., Kearney, M. and Kearney, J. M. (1997) 'IEFS pan EU survey of consumer attitudes to food, nutrition and health', *European Journal of Clinical Nutrition*, vol 51 (Suppl. 2), S1–S59

Guilbert, P. and Perrin-Escalon, H. (2004) *Baromètre Santé-Nutrition 2002*, Paris: INPES (Institut National de Prévention et d'Education pour la Santé)

London School of Hygiene and Tropical Medicine and the International Obesity Taskforce (2005) 'How much food do we really need?', Joint Symposium of the London School of Hygiene and Tropical Medicine in conjunction with the International Obesity Taskforce, 4 February, London

Prentice, A. M. and Jebb, S. A. (2003) 'Fast foods, energy density and obesity: A possible mechanistic link', *Obesity Reviews*, vol 4, no 4, pp187–194

Sweet, M. (2008) 'Children's hospital under pressure to end "grotesque" ties with McDonalds', *British Medical Journal*, vol 336, p578

Uauy, R. and Kain, J. (2002) 'The epidemiological transition: Need to incorporate obesity prevention into nutrition programmes', *Public Health Nutrition*, vol 5, no 1A, pp223–229

WHO (2000) *Obesity: Preventing and Managing the Global Epidemic*, Report of a WHO Consultation, WHO Technical Report Series 894, Geneva: WHO

WHO/FAO (2003) *Diet, Nutrition and the Prevention of Chronic Diseases*, Report of a
 WHO/FAO Expert Consultation, WHO Technical Report Series 916, Geneva: WHO
Zimmermann, M. B., Zeder, C., Muthayya, S., Winichagoon, P., Chaouki, N., Aeberli, I.
 and Hurrell, R. F. (2008) 'Adiposity in women and children from transition countries
 predicts decreased iron absorption, iron deficiency and a reduced response to iron
 fortification', *International Journal of Obesity*, vol 32, pp1098–1104

Chapter 4

Cornu, A., Massamba, J. P., Traissac, P., Simondon, F., Villeneuve, P. and Delpeuch, F. (1995)
 'Nutritional change and economic crisis in an urban Congolese community', *International
 Journal of Epidemiology*, vol 24, no 1, pp155–164
FAO (1996) 'World Food Summit 13–17 November 1996', technical background document,
 FAO, Rome
FAO (2000) *The State of Food and Agriculture 2000*, Washington DC: FAO
Franco, M., Orduñez, P., Caballero, B., Tapia Granados, J. A., Lazo, M., Bernal, J. L.,
 Guallar, E. and Cooper, R. S. (2007) 'Impact of energy intake, physical activity, and
 population-wide weight loss on cardiovascular disease and diabetes mortality in Cuba,
 1980–2005', *American Journal of Epidemiology*, vol 166, no 12, pp1374–1380
Geissler, C. (1999) 'China: The soyabean-pork dilemma', *Proceedings of the Nutrition Society*,
 vol 58, no 2, pp345–353
Jones, A. (2002) 'An environmental assessment of food supply chains: A case study on dessert
 apples', *Environment Management*, vol 30, no 4, pp560–576
Schäfer-Elinder, L. (2005) 'Obesity, hunger, and agriculture: The damaging role of subsidies',
 British Medical Journal, vol 331, pp1333–1336

Chapter 5

Finkelstein, E. A., Ruhm, C. J. and Kosa K. M. (2005) 'Economic causes and consequences
 of obesity', *Annual Review of Public Health*, vol 26, pp239–257
Ghezan, D., Mateos, M. and Viteri, L. (2002) 'Impact of supermarkets and fast-food chains
 on horticulture supply chains in Argentina', *Development Policy Review*, vol 20,
 pp389–408
Hawkes, C. (2005) 'The role of foreign direct investment in the nutrition transition', *Public
 Health Nutrition*, vol 8, no 4, pp357–365
Lang, T. and Heasman, M. (2004) *Food Wars: The Global Battle for Mouths, Minds and
 Markets*, London: Earthscan
McCann, J. C. (2005) *Maize and Grace*, Cambridge, MA: Harvard University Press
Millstone, E. and Lang, T. (2003) *The Atlas of Food: Who Eats What, Where and Why*,
 London: Earthscan
Neven, D., Reardon, T., Wang, H. L. and Chege, J. (2006) 'Supermarkets and consumers in
 Africa: The case of Nairobi, Kenya', *Journal of International Food & Agribusiness Marketing*,
 vol 8, pp103–123
Reardon, T. and Berdegué, J. (2002) 'The rapid rise of supermarkets in Latin American',
 Development Policy Review, vol 20, no 4, pp371–388

Zenk, S. N., Schulz, A. J., Hollis-Neely, T., Campbell, R. T., Holmes, N., Watkins, G., Nwankwo, R. and Odoms-Young, A. (2005) 'Fruit and vegetable intake in African Americans: Income and store characteristics', *American Journal of Preventive Medicine*, vol 29, no 1, pp1–9

Chapter 6

Bellisari, A. (2008) 'Evolutionary origins of obesity', *Obesity Reviews*, vol 9, pp165–180
Bouchard, C. (2007) 'The biological predisposition to obesity: Beyond the thrifty genotype scenario', *International Journal of Obesity*, vol 31, pp1337–1339
Byrd-Bredbenner, C. and Grasso, D. (2000) 'Trends in US prime-time television food advertising across three decades', *Nutrition and Food Science*, vol 30, no 2, pp59–66
Dibb, S. and Castell, A. (1995) *Easy to Swallow: Results from a Survey of Food Advertising on Television*, London: National Food Alliance
Goldberg, M. E. (1990) 'A quasi-experiment assessing the effectiveness of TV advertising directed to children', *Journal of Marketing Research*, vol 27, no 4, pp445–454
Horgen, K. B., Choate, M. and Brownell, K. D. (2001) 'Television food advertising. Targeting children in a toxic environment', in Singer, D. G. and Singer, J. L. (eds) *Handbook of Children and the Media*, Thousand Oaks, CA: Sage
Ledikwe, J. H., Ello-Martin, J. A. and Rolls, B. J. (2005) 'Portion sizes and the obesity epidemic', *The Journal of Nutrition*, vol 135, no 4, pp905–909
Lobstein, T. and Dibb, S. (2005) 'Evidence of a possible link between obesogenic food advertising and child overweight', *Obesity Reviews*, vol 6, no 3, pp203–208
Rogers, P. M., Fusinski, K. A., Rathod, M. A., Loiler, S. A., Pasarica, M., Shaw, M. K., Kilroy, G., Sutton, G. M., McAllister, E. J., Mashtalir. N, Gimble, J. M., Holland, T. C. and Dhurandhar, N. V. (2008) 'Human adenovirus Ad-36 induces adipogenesis via its E4 orf-1 gene', *International Journal of Obesity*, vol 32, pp397–406
Shadan, S. (2008) 'What's your fat-cell allowance?', *Nature*, no 453, p8
Spalding, K., Arner, E. A., Westermark, P. O., Bernard, S., Buchholz, B. A., Bergmann, O., Blomqvist, L., Hoffstedt, J., Näslund, E., Britton, T., Concha, H., Hassan, M., Rydén, M., Frisén, J. and Arner, P. (2008) 'Dynamics of fat cell turnover in humans', *Nature*, no 453, pp783–787
Wardle, J., Carnell, S., Haworth, C. M. A. and Plomin, R. (2008) 'Evidence for a strong genetic influence on childhood adiposity despite the force of the obesogenic environment', *American Journal of Clinical Nutrition*, vol 87, pp398–404
WHO (2000) *Obesity: Preventing and Managing the Global Epidemic*, Report of a WHO Consultation, WHO Technical Report Series 894, Geneva: WHO

Chapter 7

DiGuiseppi, C., Roberts, I. and Li, L. (1997) 'Influence of changing travel patterns on child death rates from injury: Trend analysis', *British Medical Journal*, vol 314, no 7082, pp710–713 (erratum in *British Medical Journal*, 1997, vol 314, no 7091, p1385)
Ellaway, A., Macintyre, S. and Bonnefoy, X. (2005) 'Graffiti, greenery, and obesity in adults: Secondary analysis of European cross sectional survey', *British Medical Journal*, vol 331, no 7517, pp611–612

Ferro-Luzzi, A. and Martino, L. (1996) 'Obesity and physical activity', *Ciba Foundation Symposium*, vol 201, pp207–221

Larkin, M. (2003) 'Can cities be designed to fight obesity? Urban planners and health experts work to get people up and about', *The Lancet*, vol 362, no 9389, pp1046–1047

NICE (2008) *Physical Activity and the Environment*, London: National Institute of Health and Clinical Excellence

Office of Population Censuses and Surveys (1994) *General Household Survey*, London: Her Majesty's Stationery Office

Rissanen, A. M., Heliövaara, M., Knekt, P., Reunanen, A. and Aromaa, A. (1991) 'Determinants of weight gain and overweight in adult Finns', *European Journal of Clinical Nutrition*, vol 45, no 9, pp419–430

WHO (2000) *Obesity: Preventing and Managing the Global Epidemic*, Report of a WHO Consultation, WHO Technical Report Series 894, Geneva: WHO

Chapter 8

Chicurel, M. (2000) 'Whatever happened to leptin?', *Nature*, vol 404, no 6778, pp538–540

Christensen, R., Kristensen, P. K., Bartels, E. M., Bliddal, H. and Astrup, A. (2007) 'Efficacy and safety of the weight-loss drug rimonabant: A meta-analysis of randomised trials', *The Lancet*, vol 370, pp1706–1713

Crémieux, P. Y., Buchwald, H., Shikora, S. A., Ghosh, A., Yang, H. E. and Buessing, M. (2008) 'A study on the economic impact of bariatric surgery', *Am J Manag Care*, vol 14, pp589–596

Franco, O. H., Bonneux, L., de Laet, C., Peeters, A., Steyerberg, E. W. and Mackenbach, J. P. (2004) 'The Polymeal: A more natural, safer, and probably tastier (than the Polypill) strategy to reduce cardiovascular disease by more than 75%', *British Medical Journal*, vol 329, no 7480, pp1447–1450

National Task Force on the Prevention and Treatment of Obesity (1996) 'Long-term pharmacotherapy in the management of obesity', *The Journal of the American Medical Association*, vol 276, no 23, pp1907–1915

Rand, C. S. and MacGregor, A. M. (1991) 'Successful weight loss following obesity surgery and the perceived liability of morbid obesity', *International Journal of Obesity*, vol 15, no 9, pp577–579

Rucker, D., Padwal, R., Li, S. K., Curioni, C. and Lau, D. C. W. (2007) 'Long term pharmacotherapy for obesity and overweight: Updated meta-analysis', *British Medical Journal*, vol 335, pp1194–1199

Sjostrom, L., Narbro, K., Sjostrom, C. D., Karason, K., Larsson, B., Wedel, H., Lystig, T., Sullivan, M., Bouchard, C., Carlsson, B., Bengtsson, C., Dahlgren, S., Gummesson, A., Jacobson, P., Karlsson, J., Lindroos, A. K., Lonroth, H., Naslund, I., Olbers, T., Stenlof, K., Torgerson, J., Agren, G. and Carlsson, L. M. (2007) 'Effects of bariatric surgery on mortality in Swedish obese subjects', *New England Journal of Medicine*, vol 357, pp741–752

Thuan, J. F. and Avignon, A. (2005) 'Obesity management: Attitudes and practices of French general practitioners in a region of France', *International Journal of Obesity*, vol 29, no 9, pp1100–1106

Torgerson, J. S. and Sjöström, L. (2001) 'The Swedish Obese Subjects (SOS) study –

rationale and results', *International Journal of Obesity and Related Metabolic Disorders*, vol 25, Suppl 1, ppS2–S4

Wald, N. J. and Law, M. R. (2003) 'A strategy to reduce cardiovascular disease by more than 80%', *British Medical Journal*, vol 326, no 7404, p1419, (erratum in *British Medical Journal*, 2003, vol 327, no 7415, p586 and *British Medical Journal*, 2006, vol 60, no 9, p823)

WHO (2000) *Obesity: Preventing and Managing the Global Epidemic*, Report of a WHO Consultation, WHO Technical Report Series 894, Geneva: WHO

Chapter 9

Drewnowski, A. and Specter, S. E. (2004) 'Poverty and obesity: The role of energy density and energy costs', *The American Journal of Clinical Nutrition*, vol 79, no 1, pp6–16

EPODE (Ensemble, Prévenons l'Obésité des Enfants): *www.epode.fr*

Epstein, L. H., Valoski, A., Wing, R. R. and McCurley, J. (1994) 'Ten-year outcomes of behavioral family-based treatment for childhood obesity', *Health Psychology*, vol 13, no 5, pp373–383

Jain, A. (2005) 'Treating obesity in individuals and populations', *British Medical Journal*, vol 331, pp1387–1390

Lock, K. and McKee, M. (2005) 'Will Europe's agricultural policy damage progress on cardiovascular disease?', *British Medical Journal*, vol 331, no 7510, pp188–189

McCarthy, M. (2004) 'The economics of obesity', *The Lancet*, vol 364, no 9452, pp2169–2170

PNNS (Programme National Nutrition Santé): *www.mangerbouger.fr*

Popkin, B. M. (2006) 'Global nutrition dynamics: The world is shifting rapidly toward a diet linked with noncommunicable diseases', *The American Journal of Clinical Nutrition*, vol 84, no 2, pp289–298

Singapore's New Programme to Prevent Overweight in Children: www.moe.gov.sg/education/programmes/holistic-health-framework/

WHO (2000) *Obesity: Preventing and Managing the Global Epidemic*, Report of a WHO Consultation, WHO Technical Report Series 894, Geneva: WHO

Chapter 10

Consumer International (2008) *The Junk Food Trap – Marketing UnHealthy Food to Children in Asia Pacific*, London: Consumers International

Department of Health (2007) *Foresight: Tackling Obesities – Future Choices Project*, London: Department of Health, www.foresight.gov.uk/OurWork/ActiveProjects/Obesity/Obesity.asp

French, S. (2004) 'Public health strategies for dietary change: Schools and workplaces', Symposium Modifying the Food Environment: Energy Density, Food Costs, and Portion Size, Experimental Biology Meeting, Washington DC, 19 April

Hyde, R. (2008) 'Europe battles with obesity', *The Lancet*, vol 371, no 9631, pp2160–2161

Jacobson, M. F. and Brownell, K. D. (2000) 'Small taxes on soft drinks and snack foods to promote health', *The American Journal of Public Health*, vol 90, no 6, pp854–857

Lloyd-Williams, F., O'Flaherty, M., Mwatsama, M., Birt, C., Ireland, R. and Capewell, S. (2008) 'Estimating the cardiovascular mortality burden attributable to the European Common Agricultural Policy on dietary saturated fats', *Bulletin of the World Health Organization*, vol 86, no 7, pp497–576

Slow Food: www.slowfood.com

Sustainable Development Commission (2006) *I Will if You Will: Towards Sustainable Consumption*, London: Sustainable Development Commission

Sustainable Development Commission (2008) *Green, Healthy and Fair: A Review of Government's Role in Supporting Sustainable Supermarket Food*, London: Sustainable Development Commission, www.sd-commission.org.uk/pages/supermarkets.html

Walker, P., Rhubart-Berg, P., McKenzie, S., Kelling, K. and Lawrence, R. S. (2005) 'Public health implications of meat production and consumption', *Public Health Nutrition*, vol 8, no 4, pp348–356

WHO (2000) *Obesity: Preventing and Managing the Global Epidemic*, Report of a WHO Consultation, WHO Technical Report Series 894, Geneva: WHO

WHO (2003) *Diet, Nutrition and the Prevention of Chronic Diseases*, Geneva: WHO

WHO (2004) *Global Strategy on Diet, Physical Activity and Health*, Geneva: WHO, www.who.int/dietphysicalactivity/en/

WHO (2007) A Guide for Population-based Approaches to Increasing Levels of Physical Activity, Geneva: WHO

Chapter 11

Cordain, L., Eaton, S. B., Sebastian, A., Mann, N., Lindeberg, S., Watkins, B. A., O'Keefe, J. H. and Brand-Miller, J. (2005) 'Origins and evolution of the Western diet', *American Journal of Clinical Nutrition*, vol 81, no 2, pp341–354

Davis, A., Valsecchi, C. and Fergusson, M. (2007) *Unfit for Purpose: How Car Use Fuels Climate Change and Obesity*, London: Institute for European Environmental Policy

Department of Health (2007) *Foresight: Tackling Obesities – Future Choices Project*, London: Department of Health, www.foresight.gov.uk/OurWork/ActiveProjects/Obesity/Obesity.asp

Department of Health (2008) *Healthy Weight, Healthy Lives. A Cross-government Strategy for England*, London: Department of Health

Egger, G. (2008) 'Dousing our inflammatory environment(s): Is personal carbon trading an option for reducing obesity – and climate change?', *Obesity Reviews*, vol 9, no 5, pp456–463

FAO (1996) 'World Summit, 13–17 November 1996', technical background document, FAO, Rome

Frank, L. D., Andresen, M. A. and Schmid, T. L. (2004) 'Obesity relationships with community design, physical activity, and time spent in cars', *Am J Prev Med*, vol 27, no 2, pp87–96

Griffiths, J., Hill, A., Spiby, J., Gill, M. and Stott, R. (2008) 'Ten practical actions for doctors to combat climate change', *British Medical Journal*, vol 336, p1507

IASO (International Association for the Study of Obesity): www.iaso.org

Kelly, T., Yang, W., Chen, C. S., Reynolds, K. and He, J. (2008) 'Global burden of obesity in 2005 and projections to 2030', *International Journal of Obesity*, vol 32, pp1431–1437

McMichael, A. J., Powles, J. W., Butler, C. D. and Uauy, R. (2007) 'Food, livestock production, energy, climate change, and health', *The Lancet*, vol 370, no 9594, pp1253–1263

Michaelowa, A. and Dransfeld, B. (2006) 'Greenhouse gas benefits of fighting obesity', HWWI Research Paper 4–8, Hamburg, http://www.hwwi.org/Publications_Single. 5093.0.html?&L=1&tx_wilpubdb_pi1[authorid]=49&tx_wilpubdb_pi1[singleView]= 1&tx_wilpubdb_pi1[publication_id]=291&tx_wilpubdb_pi1[back]=1068&cHash= 1b9a914bed

Myr, R. (2008) 'Breastfeeding tackles both obesity and climate change', *British Medical Journal*, vol 331, no 7510

Roberts, I. (2007) 'How the obesity epidemic is aggravating global warming', *New Scientist*, no 2610, pp165–166

Stern, N. (2006) *Stern Review on the Economics of Climate Change*, London: HM Treasury, www.hm-treasury.gov.uk/sternreview_index.htm

Sustainable Development Commission (2008) *Green, Healthy and Fair: A Review of Government's Role in Supporting Sustainable Supermarket Food*, London: Sustainable Development Commission, www.sd-commission.org.uk/pages/supermarkets.html

Woodcock, J., Banister, D., Edwards, P., Prentice, A. and Roberts, I. (2007) 'Energy and transport', *The Lancet*, vol 370, no 9592, pp1078–1088

Epilogue

Chopra, M. and Darnton-Hill, I. (2004) 'Tobacco and obesity epidemics: Not so different after all?', *British Medical Journal*, vol 328, no 7455, pp1558–1560

Godlee, F. (2005) 'Editor's choice: Untangling a skein of wool', *British Medical Journal*, vol 331, no 7510

Kersh, R. and Morone, J. (2002) 'The politics of obesity: Seven steps to government action', *Health Affairs*, vol 21, no 6, pp142–153

Yach, D., Hawkes, C., Epping-Jordan, J. E. and Galbraith, S. (2003) 'The World Health Organization's Framework Convention on Tobacco Control: Implications for global epidemics of food-related deaths and disease', *Journal of Public Health Policy*, vol 24, no 3–4, pp274–290

Further reading

General information

Blackburn, G. L. and Walker, W. A. (2005) 'Science-based solutions to obesity: What are the roles of academia, government, industry, and health care?', *The American Journal of Clinical Nutrition*, vol 82, supplement, pp207S–273S

British Medical Association (2005) *Preventing Childhood Obesity*, Board of Science, London: BMA

Campbell, P. and Dhand, R. (2000) 'Nature insight: Obesity', *Nature*, vol 404, no 6778, pp631–677

Drewnowski, A. and Rolls, B. J. (2005) 'Symposium: Modifying the Food Environment: Energy Density, Food Costs, and Portion Size', *The Journal of Nutrition*, vol 135, pp898–915

International Journal of Epidemiology (2006) 'Special themed issue: Obesity', *International Journal of Epidemiology*, vol 35, no 1, pp1–210

Kumanyika, S., Jeffery, R. W., Morabia, A., Ritenbaugh, C. and Antipatis, V. J. (2002) 'Public Health Approaches to the Prevention of Obesity (PHAPO) Working Group of the International Obesity Task Force (IOTF). Obesity prevention: The case for action', *International Journal of Obesity & Related Metabolic Disorders*, vol 26, no 3, pp425–436

Lobstein, T., Baur, L., Uauy, R. and IASO International Obesity Task Force (2004) 'Obesity in children and young people: A crisis in public health', *Obesity Reviews*, vol 5, suppl 1, pp4–104

Lobstein, T., Rigby, N. and Leach, R. (2005) *EU Platform on Diet, Physical Activity and Health*, London: IOTF/IASO/EASO

Science (2003) 'Special Issue: Obesity', *Science*, vol 299, no 5608, pp845–860

WHO (2002) *Globalization, Diets and Noncommunicable Diseases*, Geneva: WHO

WHO (2004) *Global Strategy on Diet, Physical Activity and Health*, 57th World Health Assembly, Report by the Secretariat, Geneva: WHO

Yach, D., Hawkes, C., Gould, C. L. and Hofman, K. J. (2004) 'The global burden of chronic diseases: Overcoming impediments to prevention and control', *The Journal of the American Medical Association*, vol 291, no 21, pp2616–2622

Specific topics

Portion size

Rozin, P., Kabnick, K., Pete, E., Fischler, C. and Shields, C. (2003) 'The ecology of eating: Smaller portion sizes in France than in the United States help explain the French paradox', *Psychological Science*, vol 14, no 5, pp450–454

Influence of infancy and childhood on obesity

Reilly, J. J., Armstrong, J., Dorosty, A. R., Emmett, P. M., Ness, A., Rogers, I., Steer, C., Sherriff, A. and Avon Longitudinal Study of Parents and Children Study Team (2005) 'Early life risk factors for obesity in childhood: Cohort study', *British Medical Journal*, vol 330, no 7504, p1357

Viner, R. M. and Cole, T. J. (2005) 'Adult socioeconomic, educational, social, and psychological outcomes of childhood obesity: A national birth cohort study', *British Medical Journal*, vol 330, no 7504, p1354

Double burden

Doak, C. M., Adair, L. S., Bentley, M., Monteiro, C. and Popkin, B. M. (2005) 'The dual burden household and the nutrition transition paradox', *International Journal of Obesity*, vol 29, pp129–136

Standing Committee on Nutrition of the United Nations System (2006) 'Tackling the double burden of malnutrition?', *SCN News*, vol 32, pp1–72

Obesogenic environment

Egger, G. (2008) 'Dousing our inflammatory environment(s): Is personal carbon trading an option for reducing obesity and climate change?', *Obesity Reviews,* vol 9, pp456–463

Egger, G., Swinburn, B. and Rossner, S. (2003) 'Dusting off the epidemiological triad: Could it work with obesity?', *Obesity Reviews,* vol 4, no 2, pp115–119

Swinburn, B. and Egger, G. (2002) 'Preventive strategies against weight gain and obesity', *Obesity Reviews,* vol 3, pp289–301

Swinburn, B. and Egger, G. (2004) 'The runaway weight gain train: Too many accelerators, not enough brakes', *British Medical Journal,* vol 329, no 7468, pp736–739

Swinburn, B., Egger, G. and Raza, F. (1999) 'Dissecting obesogenic environments: The development and application of a framework for identifying and prioritizing environmental interventions for obesity', *Preventive Medicine,* vol 29, pp563–570

Index